A SHEARWATER BOOK

SO SHALL

YOU REAP

ISLAND PRESS / Shearwater Books

Washington, D.C. / *Covelo, California*

SO SHALL YOU REAP

Farming and Crops in Human Affairs

Otto T. Solbrig & Dorothy J. Solbrig

A *Shearwater Book*
published by Island Press

Copyright © 1994 by Island Press
Illustrations, other than those on pages 132 and 212, copyright © by
Abigail Rorer.

LIBRARY OF CONGRESS
CATALOGING-IN-PUBLICATION DATA
Solbrig, Otto Thomas
So shall you reap : farming and crops in human affairs / Otto T.
Solbrig and Dorothy J. Solbrig.
P.   cm.
Includes bibliographical references and index.
ISBN 1-55963-308-5 (cloth). —ISBN 1-55963-309-3 (pbk.)
1. Man—Influence on nature. 2. Agriculture—History.
3. Agriculture—Environmental aspects. 4. Traditional
farming—History. I. Solbrig, Dorothy J., 1945–   . II. Title.
GF75.S68 1994
306.3'49—dc20   93-43330
                                                       CIP

Printed on recycled, acid-free paper

Manufactured in the United States of America

10 9 8 7 6 5 4 3 2

For Milton and Hannah,
whose love and good humor
are always with us

Contents

Prologue

THIS IS A BOOK about how growing plants has affected humankind. It is more than a short history of agriculture. It is about how farming evolved—when and how and why people learned to irrigate, to fertilize, and to rotate their crops. More important, it is about crops and how they connect with historical events: cereals' relation to the beginning of civilization, sugar's to slavery, the potato's to famine in Ireland. Above all it is about the link between farming and changes in the environment.

A million years ago the earth was home to only a few million human beings, all living in Africa. Today there are 5.5 billion people, all over the world, and the population is growing at the rate of some 100 million a year. A million years ago people roamed through the countryside in family groups or small bands. Today much of the population lives in complex urban centers, and those who don't are profoundly affected by what goes on in cities. People a million years ago were hunter-gatherers whose impact on the environment was about the same as that of other large animals and limited to one continent. Today much of the earth's land surface has been modified by human activity, and the pace is increasing. None of the change would have come about without agriculture.

Farming is the source of most of the food we eat, much of the fiber with which we make our clothing, the fodder for our animals, and the raw materials for our beverages. Without agriculture, our way of living would not be possible. But farming inevitably transforms the environment. Since the adoption of agriculture, people the world over have modified landscapes, destroyed forests, eliminated species, and altered ecosystems—environmental transformations that ultimately threaten human well-being and survival. When there were only a few hundred thousand people on earth, the changes caused by agriculture were tolerable. Now that there are more than five billion, such changes are becoming unbearably costly. This is not to say that landscape transformation is a phenomenon of which people have only recently become aware. Plato lamented the loss of the pine forests of Attica, and in 1582 King Philip II of Spain commented in a letter to the president of the council of Castilla, "One topic that I wish to have considered relates to the matter of the conservation of forests and their increase, since they are much diminished; I fear that those that will follow us will have much to complain [of] if we destroy them, and I pray to God that we will not see it happen in our day."[1]

People in all parts of the world and at different times have modified their environment, sometimes improving it by planting trees, protecting watersheds, and reducing erosion, sometimes degrading it by cutting down forests, overgrazing grasslands, or dumping refuse into streams and lakes. Farmers in particular have transformed natural landscapes to create the conditions necessary to grow on a large scale a few plant species and even fewer species of domesticated animals. Farmers felled primeval forests to make room for cultivated fields, quickening the pace of erosion on slopes and mountains. They drained swamps and dammed or diverted rivers to provide water for irrigation and human consumption.[2] The relation between deforestation and agriculture has long been recognized. In 1851, at the annual meeting of the British Association for the Advancement of Science, Dr. Hugh Cleghorn gave a report entitled "Probable Effects in an Economical and Physical Point of View of the Destruction of Tropical Forests": "From the number and extent of the forests and jungles of India, it might be inferred that timber was abundant in all parts, not only for home consumption, but that a supply might be obtained for foreign commerce: this is far from being the case. Though forest lands are extensive, their contents in accessible situations are not of a nature, or sufficiently abundant, to supply even the ordinary demands. In India as in other long inhabited and early civilized countries, the parts best adapted for agricultural purposes have long been cleared of jungle."

Although farming has affected landscapes all over the world, this book concentrates on the Western agricultural tradition, which originated in the Middle East, and on its modern manifestation, industrialized farming. Western agriculture, based on the use of the plow, is responsible for much environmental change. It is also the most productive type of farming and holds the promise of solving many of the world's food problems.

Farming not only affected the environment, it profoundly changed human society. Historically, the greatest change brought about by farming was social. The invention of tools and the development of agriculture transformed the human species from small nomadic bands whose activities had little ecological impact to highly complex and interrelated societies engaged in industrial activities affecting all life on the planet.[3]

People are not always aware of how deeply agriculture has shaped their way of life and their religious beliefs. If a couple who lived 11,000 years ago were transported to modern Iowa, they would be dumbfounded. She, accustomed to collecting plants wherever they grew, would not be allowed to walk into a field and help herself to the maize or soybeans growing there. Neither could he hunt cows or pigs—at least, not without being arrested and deprived of his freedom. The idea that such products of nature are private property would be completely alien to their way of thinking. They would be astonished to hear how many hours people work. In their own time the men probably did not spend more than about four or five hours a day hunting, nor did the women spend more than that gathering wild edible plants. Eleven thousand years ago people lived in camps, moving every few weeks or months to follow the supply of game and wild plants. Both isolated farms and cities with their concentrations of people living in permanent houses would seem alien to them, as would the tractors, plows, and reapers that permit single farmers to cultivate hundreds of acres. Bows, arrows, knives, scrapers, axes, and a bowl or two were all they had, and these were carried on their backs. But the most astonishing aspect of a modern farming community would be the absence of natural landscapes. Our hunter-gatherer couple would long for the forests and natural fields that provided game, wild fruit, and seeds throughout the year.

Natural landscapes must have been a fundamental part of the prehistoric world view. To this day, harmony with nature is central to the thinking of hunter-gatherers. Most likely as humans became more proficient with tools and then adopted agriculture, they increasingly saw nature as theirs for the taking.[4] Farming people tend to regard nature this way, and the notion is an integral part of the Western creation myth as set out in Genesis 1:26: "Then God said, 'Let us make man in our image, after our likeness: and let them have dominion over the fish of the sea, and over the fowl of the air, and over the cattle, and over all the earth, and over every creeping thing that creepeth upon the earth.'"

Much has changed since people began planting 10,000 years ago. Growing food has almost completely replaced gathering food. From depending entirely on their own labor, farmers have come to rely mostly on other sources of energy to sow, tend, and reap. First they used domesticated animals, then machines powered by coal and petroleum. Farmers have also moved from growing food to satisfy the needs of a single household to growing it mostly for sale to people in faraway cities. The transition is not yet complete. Many farmers in developing countries still use animal power, and about 30 to 40 percent of the world's population still grows its own food. This is changing rapidly.

Today few people are farmers, particularly in the developed world. According to the World Bank, in 1990 more than half of the world's people lived in cities, and even in the poorest countries an average of 40 percent of the population was urban.[5] There is a great deal of variation from country to country. In Uganda close to 70 percent of the population is rural, while in Singapore almost all the population is urban. However, many urban dwellers make a living from agriculture either directly or indirectly. In 1980, 4 percent of the American labor force were farmers, and they produced 3 percent of the gross national product. But farmers have to bring their products to market; produce must be prepared for storage and sold to the consumer. In 1980, besides the 3.8 million Americans employed in farming, 7.1 million sold farm products and 2 million built farm equipment and supplies, a total of roughly 13 percent of the working population.[6] Furthermore, many agricultural products are used in industries such as textiles (cotton) or paints (linseed). If we include these industries, agribusiness is the largest economic activity in the United States and in most countries.

The proportion of farmers in the population shrinks every year, yet producing, processing, distributing, packaging, and selling food employs more people than any other human occupation. What makes this industry possible is the human ability to store food for extended periods and to transport it almost anywhere in the world. The incentive that keeps the enterprise going is no longer the immediate needs of the farmer, as in early agriculture, but profit. Those who have money have access to more and better food than those who are poor.[7] It is ironic that farmers, who are generally poorer, often eat less well than urban residents. Although in developing nations there are still many subsistence farmers who do not practice industrial farming, their numbers are shrinking.

If farming profoundly altered human society, it also altered plant life. People learned to direct other species to grow in specific places and in specific ways. As humans came to depend on their crops, crop species became dependent on humans. Numerous and dramatic differences exist between cultivated plants and their wild relatives. Agriculturists selected plant varieties that were easiest to harvest or that had other characteristics desirable for farming. Through selective breeding, people lessened the ability of plants to reproduce and compete in the wild. Today, most crop plants cannot survive without human aid.

Agriculture affected not only crops but every other plant and animal species as well. When forests and savannas were transformed into agricultural fields and pastures, many species were displaced. Species such as the passenger pigeon and sandalwood that could not adapt to agricultural environments became extinct or were drastically reduced in number.[8] Those that thrived under the plow and in the environment it created multiplied and dispersed all over the world—the common rat, English sparrow, starling, housefly, cockroach, dandelion, and numerous other species.[9] Crops and domesticated animals were so transformed by human selection that they lost their ability to reproduce in the wild and are now dependent on people for their existence.

What led humans to abandon the hunting and gathering life they had followed for over a million years is unclear. Likely a number of factors were responsible, including population growth. Whether farming led to an increase in human population or vice versa no one knows. Agriculture most certainly created food surpluses that allowed populations to increase. Surpluses allowed some members of society to spend their time in specialized activities that laid the foundation for civilization, art, and industry. Industry in turn improved agriculture and made it into a vast enterprise. As food surpluses led to an increase in population, farming intensified, along with its impact on the environment. Today's response to an increase in population is to increase food production. But there is another way of solving food shortages: diminishing demand by controlling population growth.[10] This notion conflicts with the view of an infinitely plentiful nature that agricultural and pastoral societies have held for the last 6,000 years.

Humans presently farm only a small portion of the earth's surface. According to the United Nations (UN) Food and Agriculture Organization, in 1985 the world's cultivated lands comprised about 1.48

billion hectares, only 11.3 percent of the land surface of the earth. Another 3.16 billion hectares, representing 24.1 percent of the surface of the planet, were permanent pasture.[11] Forest and woodland covered 4.08 billion hectares (31.2 percent). Of the remaining 4.36 billion hectares (33.3 percent), a small proportion was devoted to urban uses and roadways. Much of the remaining land was mountainous terrain, desert, and frozen tundra. Only 5 million hectares, much of it forest, part of it desert and mountainous terrain, had been set aside as protected wilderness. These figures appear to indicate that ample land is still available to grow plants: Less than half the surface of the earth is used directly by humans. Yet not all land is suitable for agriculture. Ideal agricultural land is flat and rich, receives adequate rainfall, and has the capacity to retain water and nutrients. A more detailed look reveals that in the temperate zone all such land, and in the tropics most, is employed by humans.

World population stands today at 5.5 billion. Projections by the UN and other agencies indicate that the population will double over the next fifty to one hundred years, and that it might—and only might—stabilize after that. To avoid worldwide famine, agricultural production will have to double within that time. In the absence of appropriate new land, productivity will have to increase on existing farmland. The critical challenge will be to double production without massive environmental degradation and ultimate loss of life.

Modern agriculture is threatening the environment with nitrogenous fertilizers that seep into the water supply and with poisonous insecticides that endanger wildlife. The appropriation of forestland for agriculture is killing thousands of species of plants, animals, and microorganisms, referred to collectively as the earth's biodiversity. An increase in paddy-rice farming and in herds of ruminants has led to a greater concentration in the air of methane, a greenhouse gas that contributes to global warming. Surface erosion from ground bared by plowing removes millions of tons of topsoil every year. These are only a few of the detrimental effects of widespread agriculture. However, not to increase food production would endanger the lives of untold millions of people. The challenge is to increase agricultural production in a sustainable way so as to avoid environmental disaster. And the crucial question is how long humanity will be able to provide food for itself if its ranks continue to swell.

Only very recently in their history did humans take to growing the food they eat. For one or two million years before the adoption of farming (depending on when we consider our ancestors to have become human), people lived from the plants they gathered and the animals they hunted. If the history of mankind were to extend over a period of twenty-four hours, agriculture would exist for only the last five minutes. Yet what a difference those few minutes make! Humans have become sedentary, mostly urban creatures living on the average three to four times as long as their prehistoric ancestors. In five minutes of existence the species has increased ten-thousandfold, and most of that growth has taken place in the last ten seconds. In the twenty-three hours and fifty-five minutes that preceded the adoption of agriculture, the human population increased less than a thousandfold. The recent expansion is in part a response to the more plentiful food supply resulting from improved agricultural methods. Without modern agriculture, the more than five billion persons in the world could not be fed. And in the next hundred years humans will have to find food for at least five billion more.

To some extent, then, farming is responsible for the shape of human history. History is not so much the story of past events as it is the consciousness and interpretation of those events. In the tens of thousands of years that humans lived by hunting and gathering, they experienced droughts, storms, earthquakes, volcanic eruptions, fires, and other disturbances. They migrated to new regions, moving out of Africa into Asia and Europe and eventually America. They learned to live in new habitats—deserts, mountains, coasts, forests—and they learned how to use local resources. Cultures preserved the memory of some movements and adaptations in myths and legends. Still, nearly all people lived in the same territory as their parents, using the same skills and acquiring the same knowledge. These remote ancestors must have viewed change as imperceptibly slow, and they probably did not perceive it as history. Most change, aside from the cataclysmic sort, involved small discoveries adapted matter-of-factly into daily life. After farming societies built civilizations, change in the way people lived was increasingly the result of human actions rather than natural forces. The creation of tools, buildings, and writing produced a record, albeit incomplete, of those actions. Today we refer to the record, and our interpretation of it, as history.

Victorians would have interpreted changes in agriculture such as farmers growing food in order to buy food as a sign of progress. Whether it is depends on how progress is defined. If humans are superior to their ancestors, that is, braver, gentler, wiser, happier, and more intelligent, rational, sensible, and prudent, then the many changes they introduced can be called progress. But who can say that this is so? It would probably be more accurate to say that modern humans are more capable of functioning in a modern environment, while their remote ancestors were more capable of functioning in a prehistoric environment. In other words, the changes brought about by agricultural societies are adaptive responses to changes in their physical and social surroundings.

How do the innovations in our evolution come about and what directs them? Most likely they are the result of natural selection.[12] To survive, all species must obtain sustenance from the environment, withstand the rigors of that environment, and reproduce. No two members of the same species are identical. Within a given species some individuals are better suited to survive than others because of certain physical and behavioral characteristics. Some of these are encoded in their genes; others are the result of chance occurrences during development. Humans (and some higher vertebrates) have the ability to apply experience to increase the chance of survival and reproduction.

In each generation some individuals leave more offspring than others. They pass their genetic characteristics to their offspring. If that success is due to some characteristic provided by a gene, the gene will slowly become more widespread in the population until all the individuals in the population possess it. Biologists would say that the gene has been "selected." However, environments change, and what at one time is a beneficial characteristic can become a liability. That is why evolution cannot necessarily be read as progress.

Where the human species differs from most others is in its ability to transmit information to descendants in forms other than genes. Tradition, music, literature, photography, cinema—these allow humans to record information and make it available to future generations. Knowledge allows the species to compete and reproduce. Today education is the overwhelming factor in human dominance.

One more point needs to be mentioned. Natural selection favors the production of the largest number of offspring. Members of all species tend to reproduce as much as possible when given the opportunity. What controls the growth of animal and plant populations is the availability of resources. Whenever a new or better way of obtaining food is developed, the human population grows until it finds itself struggling to avoid starvation. In turn, population pressure pushes humans to devise new ways to increase their nourishment.[13] The pattern of expansion, crisis, invention, followed by expansion, has been repeated several times in human history.

Over the millennia, most historical change has come about almost imperceptibly. People have toiled, rejoiced, and grieved since time immemorial. When faced with problems, they have tried to solve them to the best of their ability in order to improve their lot and that of their loved ones. The result of their successes and failures is human history. With hindsight it becomes clear what people did wrong and what they did right. Our job here is to provide readers with some hindsight so that they may carry on with the work of doing what is right—or so we hope.

SO SHALL YOU REAP

Early Food Acquisition

FOR TENS OF thousands of years before the adoption of agriculture, people lived by hunting, fishing, and gathering in the wild. They ate fruits, seeds, leaves, roots, birds, snakes, frogs, snails, maggots, insects. A given band would consume over one hundred different kinds of plants and animals in a year. The composition of the diet differed from place to place and season to season. To obtain food hunter-gatherers moved from one region to another, camping for weeks or months in one place and moving to another when food got low. They probably revisited sites every year or two. For most of us such a life would be unbearable. We therefore think that it must have been miserable for our hunter-gatherer ancestors. It is of course impossible to know how people really felt back then, yet the study of earth's few remaining hunter-gatherer societies does not confirm a dismal assessment of that life-style.

Scientists have established beyond reasonable doubt that the human species originated in Africa and that it shares ancestors with apes and monkeys. The details and chronology of human ancestry, however, are far from resolved. The principal source of anthropological information is the remains of the skeletons of human forebears. Scientists have found only a few hundred prehistoric skeletons, all incomplete. Great importance is attached to these meager remains. Where people lived, the kind of food they ate, the climates they experienced, the tools they made, charcoal from their fires, scraps of animal bones and seeds in their camps—these provide additional clues to human evolution.

Extrapolating from such scanty evidence, scientists have grouped the earliest hominids, or members of the primate family to which humans belong, in the genus *Australopithecus*.[1] The oldest fossil remains of a species presumed to be ancestors of humans, A. *afarensis*, are of this genus. Anthropologists have found their bones in Kenya, east Africa, and have estimated them to be about 3.75 million years old. Fossils have also been found of several other species of *Australopithecus*, for example A. *robustus*, which lived from 2 to 1.5 million years ago in South Africa, and A. *boisei*, which existed 2.5 to 1.2 million years ago in east Africa. *Australopithecus* lived at a time when the world's climate and vegetation were changing. The rainforests that covered much of Africa were giving way to savannas, grasslands with scattered trees. *Australopithecus* probably lived where forests and savannas meet. Scientists believe that *Australopithecus* walked erect, could run quickly, and was about half the size of an average modern human. Anthropologists believe these early hominids might have been tool users, but there is no firm evidence.

Paleontologists first found fossil remains of a species presumed to descend from A. *afarensis* and be ancestral to humans in Olduvai Gorge, Kenya. They named this species *Homo habilis*, which in Latin means "skillful man," because of the large quantity of tools such as hand axes and choppers found with their bones. *Homo habilis* presumably occu-

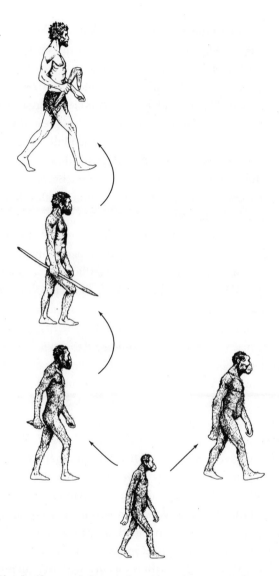

Sketch of human phylogeny, from Australopithecus afarensis *(lower figure) through* Homo habilis *and* H. erectus *to* H. sapiens *(upper left figure). Note the increase in size and the shortening of the limbs.* Homo habilis *used only crude stone tools, while* H. sapiens, *depicted here with a hoe, is associated with the beginning of agriculture. The figure on the right represents* A. robustus, *who probably shared a common ancestor with* A. afarensis.

pied the planet about 2 million years ago. Apparently they were efficient hunters, using stone tools for skinning and cutting up animal carcasses. Most but not all of the animals were small. In one case elephant bones were found near H. habilis tools. It is possible that H. habilis scavenged rather than killed large animals. Meat was clearly an important element of their diet. The presence of butchery sites suggests that these people had a home base to which they brought their kills. Their social organization might have been complex enough to allow for collaboration, at least to the extent of sharing food. There might also have been a division of labor between males, who hunted, and females, who took care of children at the home camp.

Paleontologists also found remains of a species presumed to have descended from H. habilis and to be ancestral to modern humans, Homo erectus ("upright man"), in beds dated at 1.5 million years ago or less. Homo erectus was taller and more heavily built than H. habilis and his cranial capacity was much larger. He was similar to modern humans in most physical characteristics. Homo erectus was the first humanlike species to leave Africa. Groups of them migrated to Asia. The famous Peking and Java men belong to this species. Homo erectus first stayed in tropical environments, then invaded temperate areas of China and Europe. No doubt they had to adapt physically and culturally to the colder climates.

Homo erectus learned to make complex and diverse tools such as choppers, picks, cleavers, and awls. This called for manual dexterity and an ability to plan construction. There is evidence that H. erectus used fire, probably first by taking advantage of natural fires, then by learning how to start them. The oldest human remains associated with the use of fire come from Kenya and are dated at 1.4 million years BP (before the present) and from China (700,000 BP). These might have been natural fires. The oldest hearths associated with human remains are a little under 0.5 million years old and have been found in both Europe and China. People used fire for warmth, for protection against large animals, and to cook food. Cooking tends to soften food and make it easier to chew and digest. It is interesting that about the time hearths appeared, human jaws began to get smaller and recede.[2]

Smaller and less protruding jaws are characteristic of our species, *Homo sapiens* ("wise man").

Because thousands of years separate many of these early fossils, it is difficult to establish direct continuity between "species" of humans. *Homo erectus* is undoubtedly an early form of *H. sapiens*. A strong case also can be made that *H. habilis* is a direct ancestor of modern man. It is less clear whether *A. afarensis* is a direct forerunner. He could represent an extinct branch in the tree of human ancestry.

Homo sapiens appeared about 300,000 years ago. The earliest *H. sapiens*, such as the well-known Neanderthal man, had a body like modern man's but a somewhat different skull shape. Paleontologists found remains of *H. sapiens* in Europe, North Africa, the Middle East, and Siberia. *H. sapiens* lived in France during the Ice Age and were apparently the first humans to live throughout the year in an arctic environment. *Homo sapiens* evolved into the anatomically modern human with a smaller jaw and rounder skull.

Thus hominids have roamed the planet for at least the last 2 million years. In that time they moved out of Africa and dispersed into every habitable corner of the earth. During part of the Ice Age, or Pleistocene, they had to contend with great climatic changes. Wet and dry, cold and warm periods alternated rapidly in geologic terms. In tropical latitudes the temperature during the Ice Age was not very different from what it is today, even during the coldest periods. In high latitudes, however, temperatures were lower, especially in winter. This and increased precipitation created large glaciers that covered most of Europe, Russia, and Canada. Because the ice sheets tied up great quantities of water on land in frozen form, sea levels fell and the Bering Strait that separates Alaska and Asia today became a land bridge, allowing people to walk across from Asia to America. The Mediterranean Sea separating Africa and Europe also dried up.

Hominids diverged from other primate groups after savannas had become widespread. The savanna climate is seasonal, with a five- to nine-month wet season alternating with a dry season. During the wet season plants produce many kinds of leaves, flowers, and fruits, typical primate foods. During the dry season there is little food other than roots and tubers, old leaves, and some seeds. Because of the structure of their teeth and jaws, anthropologists believe that the more robust *Australopithecus* ate large quantities of tough leaves and roots.[3] The small species of *Australopithecus*, on the other hand, ate whatever was available.

The genus *Homo*, which descends from a smaller species of *Australopithecus*, evolved along a different path. Rather than concentrating on abundant but low-quality food such as leaves and roots, they began eating larger amounts of meat. During the period (Pliocene) just before as well as during the Pleistocene era, there were many large grass-eating quadrupeds in the savannas of Africa. During the dry season they concentrated around water holes. Scientists believe that human ancestors took advantage of this situation to hunt. Hunting was riskier than eating roots, but animal food was more nutritious than plant food.

During the Pleistocene the fauna of the world included a much greater proportion than today of large mammals. Humans actively hunted animals such as bison and mammoths using a variety of techniques, from gang attacks to pit traps. Most large mammals became extinct at the end of the Pleistocene, about 12,000 years ago. One theory attributes their extinction to overhunting by humans, but their wholesale disappearance coincides with a climatic change, so the reason for it is uncertain.[4]

Some anthropologists believe that, in the late Pleistocene, humans began to hunt smaller mammals, birds, fish, and shellfish and to collect small seeds and starchy foods such as tubers.[5] For example, populations of *H. sapiens* in western Europe lived off migratory herds of reindeer.[6] Over time, their camps became larger and more numerous, and apparently they occupied them throughout the year. Remains of smaller animals such as migratory birds and aquatic species begin to show up in the refuse of camps from the late Ice Age (14,000 to 12,000 BP). This change in diet may have resulted from an increase in the human population, a decline in large-mammal populations, deterioration of the environment, or improvements in the tools and techniques for harvesting and processing food. The appearance of the remains of smaller animals accompanies an increase in grinding tools used to process tough plant materials such as seeds and roots. Seeds and tubers belonged to plants most suitable for domestication and most likely to lead to cultivation.

During the 2 million years that elapsed between their appearance and the invention of agriculture, humans developed a system of subsistence known as hunting and gathering.[7] Their diet consisted of plant materials supplemented by the flesh of wild animals. This way of life, universal 12,000 years ago, survived until recently in a few pockets, such as among the !Kung Bushmen of the Kalahari Desert, the Australian aborigines, some Amazonian Indians, and the Inuit. Thanks to that circumstance it has been possible to study this form of food procurement in detail.[8]

The few remaining hunter-gatherers, it is presumed, behave similarly to hunter-gatherers who lived 12,000 years ago. But there are many differences in particular aspects of life and social organization among the Inuit, !Kung Bushmen, and Australian aborigines. In fact, the most important lesson derived from studies of extant hunter-gatherer tribes is that they share very few characteristics. Thus extrapolations from the present to the past should be undertaken cautiously. It must be remembered, for example, that the hunting-gathering way of life survived in these groups because it was successful. This does not mean that it thrived in groups of people who abandoned it in favor of agriculture. Present-day hunter-gatherers, moreover, have been in contact with agriculturists and many have borrowed customs and tools from them.

Another important point to consider is that neither human populations nor their surroundings were static during the 2 million years that preceded the adoption of agriculture. People experienced the climatic fluctuations associated with glacial epochs. They perfected the use and manufacture of stone, bone, and wood tools, discovered the use of fire, and learned how to make baskets. These and other changes must have had profound effects on their way of life.

A prevalent notion espoused until recently by anthropologists is that the life of hunter-gatherers was fraught with danger and haunted by the specter of famine.[9] According to this view people had a very short life span and were constantly wandering in search of food. If present-day hunter-gatherers represent in some respects the life of people 12,000 years ago, nothing could be farther from the truth. Today's hunter-gatherers eat well and have plenty of leisure time, and although they move around, their stay in any given place can be lengthy. They spend only a third of their time in the pursuit of food.

The !Kung Bushmen in Botswana and neighboring Namibia are a good example of living hunter-gatherers.[10] In 1960, Harvard anthropologist Irven DeVore conducted a careful study of the !Kung that changed many preconceived notions about the life of traditional people. The Kalahari Desert is one of the most inhospitable regions of the world, and agriculture without irrigation is not feasible. The !Kung form groups of twenty to forty persons who assemble in camps around water holes during the dry season and radiate out from there during the rainy season. Their density is forty-one persons per hundred square miles, very low by present standards. However, it is surprisingly high for the harsh environment in which they live, and probably higher than the average human density in pre-agricultural days. The !Kung move their camps at least five times during the year, but they do not go far. Rarely do they move more than twelve miles from a waterhole. Although they are not sedentary, neither are they constantly on the move.

Men and women spend about the same amount of time obtaining food, two to three days a week, working on average not more than six hours a day for a total of twelve to nineteen hours a week. That is a short work week, one that most of us would envy. More surprising still, only married persons (women marry at about fifteen, men at twenty) obtain food for the band. Because the young and the old together account for about half the population, life among the !Kung is restful.

Little food is stored. The camp seldom has more than a three-day supply. Hunting and gathering is not organized, but groups of people go out daily in pursuit of food, which is then shared by the entire household. Men and women do, however, have well-defined roles in the provision of the camp. Women gather seeds and other vegetable matter; men do some gathering but spend most of their time hunting. Women provide two to three times more food by weight than men. Meat is valued more than vegetable food (as in modern societies), and hunting is more prestigious than gathering. The *mongongo* nut provides half the plant food consumed by the !Kung. The average daily per capita consumption of three hundred nuts weighs only 7.5 ounces but yields 1,260 calories and 56 grams of protein. Bushmen consume eighty-four other species of food plants, including twenty-nine species of fruits, berries, and melons and thirty species of roots and bulbs. They collect wood for cooking and use various plants for weaving and construction.

What do !Kung do the rest of the time? They spend about one to two hours a day cooking and eating. The rest of the day is passed weaving, building, or doing embroidery, visiting other camps, and entertaining visitors. Men spend a substantial amount of time trance dancing, which can go on all night and has religious overtones.

The Ache until very recently were full nomadic hunter-gatherers in the primary rainforests of eastern Paraguay. In small bands of ten to one hundred people, they move their campsites frequently while hunting and gathering, differing from the !Kung in this respect. On the average they eat an astonishing 3,700 calories a day (an active adult American consumes on the average only 2,700 calories per day). In contrast to the !Kung diet, more than half these calories come from hunting and less than half from gathering (about 18 percent come from honey). All members of the group share food. Meat is divided evenly among all members of the band except the hunter, who almost never eats from his own kill.[11] However, the husband, children, and siblings of a woman consume more of what she gathers than the rest of the group do.

The Ache eat more meat than the !Kung because they work more hours. Men spend close to seven hours a day hunting and preparing food, and about half an hour working on their hunting weapons. This leaves them about four and a half hours a day for socializing. Women spend on average two hours a day gathering, another two hours moving camp, and about eight hours in light work and child care.

Thus the life-styles of the !Kung and the Ache and of other contemporary hunter-gatherers such as the Australian aborigines do not conform to the stereotype of the pre-agricultural band of semistarving people constantly on the move.[12] Clearly hunter-gatherers of old were not marginal populations on the point of starvation. Had that been the case the human species probably would not have flourished over the last 2 million years. During this period humans acquired a wide range of sophisticated tools and skills. Starving people spending all their time scrounging for food most likely would not have been so creative.

And, as today, marked differences must have existed among groups in the security of their food supply and in their dietary habits. There is ample evidence indicating that hunter-gatherers specialized in the way they obtained their food and in the kind of food they ate. Those situated along the coast, especially if it was a dry and barren one such as Peru's, became fishermen. Others were mainly seed collectors. The original inhabitants of California specialized in acorn gathering; the inhabitants of northern Chile, in collecting the fruits of the mesquite tree. Some were primarily hunters, such as the American Plains Indians, who pursued the buffalo. [13]

Some groups were well off because of a reliable and nutritious food source, while others lived more precariously. The latter would have had greater incentive to devise ways to improve the quality and reliability of their food supply. Although the !Kung and the Ache do not store food, living instead from what they collect and hunt daily, in the past there were hunter-gatherers who by necessity made provision for the winter. Examples are people who lived in high latitudes, such as the Onas of Tierra del Fuego or the Inuit of the Arctic, or in regions with pronounced dry seasons such as deserts. Only after learning to store food could people permanently occupy marginal areas such as deserts or the Arctic.

Hunter-gatherers eat a great diversity of food that provides them with vital calories, vitamins, and amino acids. Bushmen were recorded as consuming 58 different species of plants in a period of a month. [14] North American Indians are known to have eaten 1,112 species of plants belonging to 120 different families, while Australian aborigines consume over 400 such species. [15] Diversity of diet is characteristic of nomadic people everywhere. However, it is unlikely that all hunter-gatherers in the past had as diverse and balanced a diet as the !Kung or the Ache. Even the !Kung rely on one species, the *mongongo* nut, for half of their caloric intake. People that depended on one or a few plant species, especially if game was scarce, could have experienced dietary shortages leading to disease. This would have been more likely in marginal environments such as deserts or dry mountains.[16] By examining the skeletons of ancient people it is possible to detect how well nourished they were. Studies suggest that some ancient hunter-gatherers had nutritional problems. Apparently men were often better nourished than women and children, suggesting that access to food was not always equal in prehistoric times.[17]

Nevertheless, disease, violence, and accidents were probably a greater risk to life in pre-agricultural days than hunger. There must have been times when famine wiped out an entire band of hunter-gatherers, but that was probably the exception and not the rule.

With the adoption of agriculture 10,000 years ago, humans gained greater control over food production at the cost of reduced food diversity. Only a few plants were, or could be, domesticated. For a long time, agriculture was practiced along with hunting and gathering. But gradually agricultural activity expanded and people obtained most of their food from farming. Today almost all people, even in developing countries, eat only farm-grown plants and domesticated animal products.

Demographers and anthropologists often discuss the link between food and population growth. Throughout most of its history, including the hunter-gatherer epoch, the human population was stable or grew slowly. From the end of the Pleistocene age about 14,000 years ago until the start of this century, the human population increased slowly. In that period world population went from around 10 million people to about 1 billion. Since then the population has quintupled, and before it stabilizes toward the end of the twenty-first century it will double again. The reason for the rapid increase in this century is that death rates started decreasing significantly, while birth rates decreased only slowly.

Death rates in Europe started to decline long before modern medicine came into being. This presents a problem of interpretation. Two general theories have been proposed to account for the decline in death rates in Europe during the last century. One attributes the improvement to changes in living conditions and sanitation. The other attributes the changes to an increase in food production. According to the latter hypothesis, before 1800 European populations suffered from chronic malnutrition because of insufficient food.[18] Changes in agricultural practices, including the introduction of new crops from the Americas such as maize and potatoes, increased food supplies in the eighteenth and nineteenth centuries. This reduced malnutrition and disease, particularly infant mortality. Population growth may be linked to nutrition.

That there is a link between disease and malnutrition is well established. Less certain is whether there was an overall food shortage in pre-industrial Europe, although periodic famines were well documented. If there was a food shortage, it was not due to a shortage of land. In periods of population expansion agricultural areas increased, and in times of population decline, such as during outbreaks of bubonic plague, land was abandoned. If indeed a food shortage existed, it must have been because food was not distributed evenly. Social factors such as limited access to arable land or excessive production costs could have affected food supply.[19]

In *Stone Age Economics* the American anthropologist Marshall David Sahlins calls hunter-gatherers the true affluent society. Although life was not devoid of risk from attacks by wild animals, the ravages of weather, and occasional food shortages, normally there was enough security and food for everybody. Individual acts of aggression did occur, but intergroup aggression was rare and organized warfare unknown. Leadership was shared and every member of a band was essentially equal to every other. The environmental impact of hunting-gathering was minimal. Why did humans abandon a sustainable way of life that had served them so well for 2 million years?

The hunter-gatherer life was as close to life in the Garden of Eden as humans have come. In adopting agriculture people increased their power over nature. Increased food supplies allowed greater human concentrations that in turn created the conditions for warfare as well as for civilization. Were planting and herding perhaps what Adam learned from the tree of knowledge?

CHAPTER

From Hunter-Gatherers
to Farmers

SOME 10,000 YEARS ago, people in various places around the world began to grow food plants and domesticate animals. This was a revolution in the history of humankind. Farming made a sedentary life possible, encouraged work specialization, narrowed the diet, increased social distinctions, and allowed a dramatic increase in human population. In other words, farming laid the ground for the development of civilization. At the same time agriculture and livestock raising changed the environment as farmers turned forests and meadows into fields and pastures.

Why didn't humans adopt agriculture earlier? Contrary to popular belief, they probably knew how to cultivate plants. Contemporary hunter-gatherers are well aware that seeds sown in favorable soil grow into plants. Living close to nature makes them careful and astute observers of the behavior of plants and animals. It is unlikely that ignorance kept our nomadic ancestors from taking up agriculture.

The common notion is that farming is safer than hunting and gathering—that once people learned to farm, they preferred cultivating to gathering plants from the wild. After all, modern farming produces a steady and reliable supply of food, and yields of cultivated plants are much higher than those of wild plants. Agriculture appears to require less work than gathering from the wild. Today's elaborate technology operating by virtue of huge capital investments has reduced risk so that farmers can produce a steady and reliable food stock for city markets. Abundant energy supplies result in high yields and reduce the amount of human labor needed to produce a unit of food. Such was not the case in Neolithic times, when agriculture was first practiced. Primitive farming required more labor than gathering, which simply involved the harvest of already-ripe fruits, seeds, leaves, roots, and tubers. Planting called for preparing soil, tending plants as they grew, harvesting them, and in some cases storing them. Dependence on the produce of a single field, moreover, increased the risk of losing the harvest to bad weather or pests. As we have seen, studies of contemporary hunter-gatherers show that they eat better and experience fewer food shortages than primitive agriculturists, and that they enjoy more leisure time.[1] Why did humankind give up its nomadic life in favor of sedentary agriculture?

Humans adopted agriculture at different times in different areas of the world. This is clear from the remains of isolated prehistoric villages occupied year-round, primitive farming tools, and evidence of the crops themselves—charred seeds, an occasional dried leaf, tubers and stems. Against the backdrop of human history, agriculture is a recent development: People started planting crops only 7,000 to 10,000 years ago. Agriculture was independently adopted in the Middle East, China, Mesoamerica, and Peru, and it might also have been an independent development in Africa, Southeast Asia, and tropical lowland South America.[2] Such autonomous origins have intrigued anthropologists, historians, and botanists. Why would human groups who had carried on a nomadic hunting-gathering existence for almost two million years in different parts of the world turn to a sedentary life of farming? Anthropologists no longer accept the explanation that farming provided people with more food and less toil. The first farmers, equipped with simple tools and limited knowledge, rode a fortune's wheel of unpredictable factors: bad weather, infestations of pests, outbreaks of plant disease, and other catastrophes. The answer may never be clear. Furthermore, because the origins of agriculture are so diverse, there may be more than one explanation.

If farming does not release people from work or the threat of hunger, it does increase what the farmer produces per unit area. On the same area of land, a farming society can feed more people than a hunter-gatherer society. A simple explanation for the adoption of agriculture is population increase in a society that cannot expand its territory. As the human population in an area expands, land becomes scarce, and gathering fails to yield sufficient food, people can feed themselves by planting.[3]

Farming, especially of seed crops, also allows people to live in less hospitable environments such as semideserts and mountains. In these areas wild plants provide nourishment for only part of the year, eliminating the possibility of permanent occupation by hunter-gatherers. However, a farmer might grow enough crops to sustain a family year-round. Many of the oldest archaeological sites with evidence of agriculture are located in subtropical mountainous areas with dry climates. It might be that because a population was growing, people moved into previously unoccupied marginal regions and developed agriculture as a way of coping with less favorable surroundings.[4]

Another hypothesis is that the adoption of agriculture was a response to changes in local climatic conditions. There is some evidence of a global drying trend in the era during which humans took up agriculture.[5] Climate change may have reduced the number of harvestable fruits and seeds and limited game, forcing people to supplement their diet with seeds from grown plants. According to this theory, as the climate grew drier people moved into riverine areas such as the valleys of the Nile, Jordan, Euphrates, and Tigris. As more people squeezed into smaller areas, they had to augment their gathered diet by planting.[6]

Botanists have proposed a complementary hypothesis.[7] They point out that most of the wild relatives of our crops grow best in disturbed sites with little vegetation. Here during part of the year drought, cold, or environmental upheavals such as landslides retard plant growth. Plants that grow in these sites are called pioneer plants—they colonize areas that will eventually be taken over by more permanent vegetation, but not until they, the pioneers, have produced a seed crop. To survive, pioneer species must colonize quickly; thus they have more seeds than related nonpioneer species. According to the theory, when people occupied marginal habitats they added pioneer plants to their diets. They learned that the seeds were nutritious and easy to store, and therefore excellent reserve provisions for the dry or cold months of the year. Yet the availability of pioneer plants was erratic, depending on the amount of disturbed land. Through the artificial creation of disturbed sites for such plants to grow, and later through the deliberate planting of seeds in these sites, farming might have come into being. In other words, according to this theory, agricultural fields were created to encourage "weedy" plants and keep more permanent vegetation out. Only later did people sow seeds and even later tend crops.

A related hypothesis for the origins of agriculture is serendipity, or the "rubbish-heap hypothesis."[8] Our forebears threw their refuse into heaps outside their dwellings. "Kitchen middens" have provided anthropologists with important evidence regarding the food habits of ancient peoples. According to the rubbish-heap hypothesis, the ancestors of our crop plants were mainly pioneer plants that colonized the disturbed sites of dumps, providing people with food plants in the immediate vicinity of their camps. The benefit of not having to walk great distances to find disturbed sites induced people to remove weeds and scratch the soil surface around their camps; in time they were deliberately sowing seeds.

Some anthropologists feel that in searching for an explanation, too much emphasis is placed on external factors such as climate and not enough on human social structure. They point out that agriculture allowed the production of surplus food with which to maintain a class of artisans who could produce such items as tools and pottery. According to this hypothesis, certain individuals in a group, "big men," convinced or coerced others to produce surpluses through planting to support the activities of people like artisans. Competition among big men led them to extract increasingly greater surpluses from the population.[9]

Another hypothesis has to do with trade. There is good evidence that Neolithic hunter-gatherers engaged in trade.[10] Many artifacts uncovered at archaeological sites came from hundreds of miles away. As humans began using tools, the need for special types of hard stone such as obsidian increased. Moreover, from early on people appreciated ornaments made from precious stones, which are typical of objects found in graves far from their origins. What would people living in semidesert or mountainous areas have to trade? One suggestion is food, specifically the hard seeds of grasses that grew in their regions.[11] According to this view, the desire to trade motivated local populations to increase the supply of seeds that stored well. Supporters of this hypothesis point out that early agricultural settlements lie along ancient trading routes. This and the "big men" hypothesis may be different aspects of the same social process.

All these hypotheses have one flaw. Dump heaps, marginal areas, and weedy plants existed for tens of thousands of years before people adopted agriculture, and in areas where people did not adopt agriculture. Not just one but many climatic changes occurred over time. Furthermore, neighboring peoples must have exchanged tools and other objects early on. Why did agriculture originate only in five separate sites around the globe within a span of just 2,000 to 3,000 years?

Anthropologists think that human groups could not adopt agriculture until they had reached a certain level of social and cultural development. The rudimentary stages of human cultural development are based on the types of tools people had. The groups that adopted agriculture were all in a similar stage of cultural development, usually referred to as the late Mesolithic or early Neolithic.[12] Certain tools such as hoes are a prerequisite for creating the disturbed environments that enable crops to sprout and grow. Also needed are baskets or similar lightweight containers to harvest seeds and permanent receptacles such as clay vessels to store the harvest. The lack of some or all of these items may explain why hunter-gatherers did not adopt agriculture earlier, even when ecological and climatic conditions were favorable.

Another requirement for the successful development of agriculture was the presence of wild plants and animals suitable for domestication. Humans domesticated most modern crops early in the history of agriculture, most in a few geographical areas, the so-called centers of origin or hearths. The presence of suitable wild plants and animals may well have spurred the adoption of planting and herding.

Agriculture, particularly the part-time activity that early farmers probably practiced, did not require permanent settlement. As long as people visited their fields regularly at sowing and harvest time, and occasionally in between, they could pursue a mixed hunting-gathering and farming life. On the other hand, farming allowed a sedentary life not possible with hunting and gathering. People adopting farming might have perceived advantages of a sedentary life that pushed them into adopting agriculture as their principal form of subsistence.[13]

The first and most obvious advantage of living permanently near the field was protecting the crop from predators, especially grazing animals. It is probably no coincidence that, except in the Americas, the adoption of agriculture took place about the time that humans domesticated grazing animals—cows, sheep, goats, pigs, donkeys, and horses. These animals and their wild relatives, attracted to fields with edible grasses, would have represented a threat to crops. Permanent settlement allowed the farmer to keep animals out of his fields. It also allowed the accumulation of implements such as hoes critical to farming. Nomadic people, especially those without pack animals, have few personal possessions. As soldiers and hoboes know very well, the value of heavy articles diminishes quickly when one is on the march.

Other advantages of sedentary life were socialization and control over the family. The life-style of contemporary hunter-gatherers requires them to live in small bands or family groups.[14] Otherwise they quickly use up all the food growing in their surroundings. Still, bands assemble at regular or irregular intervals for funerals or other ceremonies, and studies show that once gathered, they do not like to part. Permanent agriculture in settled villages allows human groups to coexist and interact. To the extent that early people welcomed the opportunity for socializing, it could have been a force leading to the adoption of continuous farming in one place.

Living in permanent settlements increased discipline in the family and improved the quality of childrearing. With men out hunting and women gathering food, neither could have exercised much control over the behavior of the other. Furthermore, child bearing and rearing were severely hampered in nomadic society. Parents had to carry small children until they were large enough to walk. Births would have had to be spaced widely. Large families were a burden to the hunter-gatherer. This is not the case with sedentary farmers, whose children can stay at home and when they reach a certain size help with the work of the farm.

Most of these hypotheses probably contain some element of truth. After the retreat of Pleistocene glaciers around 12,000 years ago, there was a warming trend that particularly affected those areas such as the Middle East, northern China, and Mesoamerica where agriculture began. People there had advanced in their ability to make and use stone tools. With the retreat of the ice, certain large mammals such as mammoths and bison on which they had largely relied became scarce. Vegetation also changed, and people probably started depending more on gathered plant products and less on game. There also may have been an increase in population. [15] Different human groups had established regular trade. At some point, for some or all the reasons given above, and for other unknown ones, people in the Middle East, Southeast Asia, northern China, Africa, Mesoamerica, and Peru started to cultivate plants. Concurrently or shortly thereafter, they adopted a sedentary way of life, thereby profoundly altering the history and appearance of the planet.

Agriculture would permit an increase in the population of the world. It would lead to the development of cities and city-states, to the growth of civilizations, and eventually to the industrial and technological world we know today. The introduction of agriculture was a true revolution in history, although no one at the time would have been conscious of it. In the words of William McNeill, agriculture

constituted perhaps the most basic of all human revolutions. Certainly the whole history of civilized mankind depended on the enlargement of the human food supply through agriculture and the domestication of animals. The costs were real, however, for the tedious labor of tilling the fields was a poor substitute for the fierce joys, sharp exertions, and instinctive satisfactions of the hunt. The human exercise of power thus early showed its profoundly double-edged character; for a farming folk enlarged dominion over nature, and liberation from earlier limits upon food supply, meant also an unremitting enslavement to seed, soil, and season.[16]

It was a revolution that was a long time taking hold. For centuries and even millennia after some people adopted agriculture, many others across the globe continued to follow a hunter-gatherer way of life. Furthermore, it took thousands of years before any population replaced gathering and hunting by planting and herding as the sole way of obtaining food. For example, in the Middle East agriculture goes back about 10,000 years, but it would take 4,000 years before a society totally dependent on agriculture would evolve there. Only in the last 200 years have agriculturists come to dominate the planet. If agriculture was advantageous, why did it take this long for people to rely solely on farming to meet their food needs?

The answer lies in what economists call the learning curve. We all have tried at one time or another to play a musical instrument or to learn a game of skill such as golf. Proficiency comes only with practice, and practice takes time. Furthermore, most skills are acquired through being taught. Imagine that you want to learn how to play the guitar, but that first you must make your own guitar, and when it is made you must teach yourself how to play. You face the same situation that early agriculturists faced.

Perhaps some young woman somewhere started dabbling by planting a few seeds near her camp. She learned that the plants she had sown produced seeds. She liked the benefit of not having to walk so far to obtain food and started experimenting with planting. Soon she found out that occasionally many plants grew, at other times none. She grew curious and tried to find out why. Slowly she learned that it made a difference whether she planted old seeds or recently harvested seeds, that seeds planted at certain times of the year yielded more than at other times, that her results improved if she planted after a rain, and so on. She probably spent her short life exploring and learning but contributing little to the food supply of the band. Her niece got interested in these planting trials. Because she had already learned from her aunt, she enjoyed greater success. Her investigations coincided with a time of food shortage. Other members of the band became inquisitive, and more people started supplementing their hunting and gathering with planting. Eventually other bands learned of these new customs and got interested. Yields were low, and there was no question of abandoning hunting and gathering. As time went by people started selecting plants more conducive to the soil. Yields went up. As planting slowly became a significant source of food, the band paid more attention to it than to gathering until eventually people felt secure enough in their knowledge of agriculture to abandon the nomadic hunting-gathering life. The greater food supply may have reduced child mortality and increased population. This would have multiplied the need for food, and so on.

Studies of the origins of agriculture rely on a diversity of data from various academic disciplines. Botanists and zoologists play an important role in identifying remains of plants and animals found in archaeological excavations and relating them to living crops. Modern genetics has developed techniques that allow us to identify possible ancestors of modern crops and domesticated animals. Geographers, palynologists, and climatologists provide information on climate and vegetation changes in the past. Yet archaeology plays the most important role.

Until recently, archaeologists and anthropologists were preoccupied with such aspects of human culture as art, architecture, and social organization. Most recovered artifacts were made of durable materials—metal, stone, or bone, for example. It also used to be that archaeologists recovered only artifacts large enough to be seen with the naked eye. This is no longer the case.

Some thirty years ago archaeologists began to use more precise techniques to salvage microscopic or near-microscopic materials. Their methods consist of carefully sifting the soil from excavations to recover even the smallest remains. The procedure is roughly as follows. First all the soil from an excavation is sifted through different-sized screens to recover small objects. Then the fine materials that pass through the sieves are mixed with water. Small slivers of wood, pieces of cloth, and plant materials that float to the surface are recovered. This procedure can produce tiny plant remains like sticks, fragments of seeds, even pieces of leaves. Sometimes archaeologists find complete seeds, but more often only fragments or charred remains of seeds. They send the organic fragments to specialists for analysis. Botanists have learned how to identify fragments using the microscopic features of hairs and cells.

Today's archaeologists are veritable sleuths applying skills worthy of the FBI. For example, they can analyze the fossilized feces ("coprolites") of humans and animals that lived hundreds of thousands of years ago to learn what they ate the previous day. Nonetheless, the record of what people ate and cultivated is still woefully incomplete. The further back in time we go, the less likely it is that we will find even a small piece of seed.

Charred seeds are the most common plant fragment recovered. Fire transforms seeds into charcoal, but most maintain their shape and even microscopic characteristics remarkably well. In dry environments such as Egypt or some caves in the Levant, archaeologists occasionally find mummified seeds, leaves, fruits, and even flowers. Sometimes scientists recover plant remains from bogs. In one case it was learned what some people buried in ancient bogs in Denmark had for lunch after experts examined their stomach contents.

Other sources of information include depictions of plants and farming activities in tombs and on tablets, and impressions of seeds in clay pots and bricks. Linguists supply information by comparing names given to crops and wild plants in different parts of the world. And scientists can now date organic remains using radiocarbon techniques. In spite of these and other tools, evidence for the origins of agriculture is still fragmentary. There is still a great deal of informed guesswork and sleuthing involved. New data are obtained all the time and with them our vision is constantly being updated.

Another way to study earliest agriculture is to find out where crops originated. In the 1920s the Russian botanist Nicolai I. Vavilov said it might be possible to identify where a crop was first domesticated by studying the abundance and dissimilarity of crop varieties (called land races) cultivated locally today.[17] In Vavilov's estimation, the place with the largest number of land races for a crop was the area where it had originated, since here farmers would have had the most time to modify the original wild plant. After studying the variation of land races for a large number of crops assembled in the Genetics Institute of Leningrad, Vavilov concluded that most cultivated plants had originated in only a few areas. These he called centers of origin, but a better name is centers of variation. Vavilov proposed five major centers. The first was southwest Asia, including India, southern Afghanistan, and the adjacent areas of Cashmere, Iran, Asia Minor, and the Caucasus. This was the region where soft wheats, rye, small-seeded flax, lentils, broad beans, chickpeas, a variety of vegetables, and Old World cottons were first domesticated. The second was Southeast Asia, that is, the mountainous areas of China, Tibet, Japan, Nepal, and neighboring areas. Here, according to Vavilov, barley was domesticated, along with naked oats, millets, and soybeans; crucifers such as cabbages, rape, and radishes; and many fruit trees (plums, cherries, peaches, and so on). The third center, comprising the lands around the Mediterranean Sea, was the home of hard wheats, cultivated oats, large-seeded flax, peas, large-seeded lentils, beets, and other vegetables. The fourth center was Ethiopia and adjacent areas, where a large diversity of varieties of cultivated plants exist. This was supposedly the center of origin for some forms of barley and wheat, sorghum, and coffee. The fifth and final center was in the Americas, specifically the highlands of Mexico and Peru, where inhabitants domesticated maize, potatoes, beans, tobacco, Jerusalem artichokes, and American cotton. Three of Vavilov's centers (southwest and Southeast Asia, Mexico, and Peru) are places for which we have firm archaeological evidence that people took up agriculture independently. The inhabitants of the Mediterranean area learned of agriculture from the Middle East, but they did so relatively early. Similarly, the Ethiopian people did not take up agriculture independently, but they acquired it early on.

Today the theory is no longer tenable that where a crop is most variable signifies its place of origin. Local topography, climate, and history are factors that can increase diversity. A good example is Ethiopia. This rugged, geographically diverse area with its variable climate and long history of occupation by traditional farmers has produced a great diversity of crop varieties. Yet there is no good evidence that Ethiopians were the first people to domesticate the wheat, oats, and barley they grow. On the contrary, there is some indication that the Cushites, a group of people of Caucasoid origin and Hamitic stock, introduced several of these varieties.[18] Another example is Hungary. While many different kinds of red peppers are used to prepare paprika there, the red pepper is an American plant that Hungarians could not have grown before the time of Columbus.

A variety of archaeological, genetic, botanical, and geographical studies have verified the hypothesis that most crops have been domesticated in a limited number of places. Four of these (China, the Middle East, Mesoamerica, and Peru) are areas where evidence unmistakably indicates that people adopted agriculture independently; others (such as Europe, India, Southeast Asia, Oceania, Africa, the southwestern United States, and tropical South America) are areas where it is more likely that people domesticated new crops after agriculture had been introduced. They are often called noncenters. Experts differ over the number and geographical extent of noncenters.[19]

Most likely, the first agriculturists were unaware that they had developed a new way of obtaining food. They must have seen themselves as improving on their age-old system of acquiring food. The original farmers were probably women, just as the initial herders were probably men. Women do most of the plant collecting in hunter-gatherer societies, while men do the hunting. Herding and planting can be considered as extensions of these activities.

Where there is sufficient archaeological evidence, as in the Middle East and Mexico's Tehuacán Valley, we can see that many centuries elapsed before agriculture became the dominant life-style. People might have been reluctant to increase the labor devoted to farming, and food from wild plants and animals was sufficient to satisfy needs. Eventually larger settlements and complex civilizations based on the activity of farming arose in both areas. Are these developments the inevitable outcome of an agricultural way of life? What induced people to accept the toil and the discipline associated with farming?

Humans did not choose agriculture with the express purpose of feeding more people per area. Such purposeful behavior does not agree with most readings of human history.[20] However, once people adopted agriculture and a sedentary life, even if only partially, they discovered that more people could be fed. And the more food that was available, the more people reproduced. Demographers estimate that the population in the Middle East multiplied about sixteen times in the period between 8000 and 6000 BP.[21]

No one can determine with certainty whether a people adopted agriculture because the population was growing, or whether the number of people increased because there was more food available. The latter view tends to prevail, but the process could not have been so clear-cut. The availability of food, as well as the change to a sedentary life, supported an increase in the population. As the population grew, local supplies of wild plants and game became depleted, requiring increased supplements from farming and herding. The repetition of this cycle eventually forced the population into agriculture as a permanent way of life. To this day, traditional farmers adjust their output to the needs of their families; as needs decline, so too does the area under cultivation and the time the farmer works.[22]

As populations increased and civilizations arose, several fundamental changes took place in the farming way of life. Very soon after agriculture was embraced in the Middle East, household goods increased along with polished stone tools, especially the axes and sickles that are a hallmark of Neolithic agricultural settlements. While rough-hewn flint tools, chipped and flaked to produce sharp edges, made adequate arrowheads and scrapers for the hunt, they were not useful for cutting down trees. The first remnants of cloth also coincide with the first evidences of farming. Basket and pottery making might have predated agriculture, but until there was a reliable supply of plant and animal fibers woven cloth could not replace animal skins as the basic form of apparel. Domesticated sheep and goats and cultivated flax enlarged the supply of fibers for spinning and weaving.

Once farming and herding became a way of life, the cycle of seasons assumed an importance unmatched in hunting-gathering society. Farmers have to prepare the ground before sowing and plant at precise times of the year. They have to protect their crops from grazing animals and cultivate them to keep out weeds. They have to harvest their crops at precisely the right time or they risk losing them. If farmers harvest crops too early, when the seeds are not ripe, they are likely to rot in winter. If too much time elapses before reaping, seeds ripen and fall off—and most fallen seeds are lost. Furthermore, farmers have to assess how much land they need to plant to feed the household. They must calculate the amount of seed needed for next year's sowing and put it aside. Careless or lazy farmers who fail in these chores may not grow sufficient food for their families. Most hunter-gatherers share food freely, but sharing is not so free when family supplies must be carefully calculated.

If a single crop is farmed continuously, soil soon becomes exhausted and crop yields diminish. Then it is time to move to another place or to rotate the crop. Early farmers were probably seminomadic, as frontier farmers usually are. Only after populations increased and suitable unoccupied land became scarcer would farmers have learned about the importance of crop rotation and the beneficial effect of animal grazing on soil fertility.

The first communities in the Middle East, China, and the Americas to practice agriculture did so by adding artificially raised crops to a hunting and gathering food economy. This early agriculture was simple. It consisted of scratching the surface with stone axes or primitive hoes, sowing, and reaping. Early farmers cultivated their fields until the soil became exhausted, probably not for more than two or three seasons. Then they planted new fields. When there was no virgin land left, farmers cultivated fields that had lain fallow the longest. Yields must have been low. The principal advantage of agriculture at that stage was producing plants more useful than the native vegetation. This type of cultivation, called swidden, shifting, or slash and burn, was practiced until recently over large parts of the tropics and still survives in some areas.

Shifting cultivation is most prominent in wooded areas. Farmers clear an area, most often partially, of trees and shrubs, allow the debris to dry, and then burn it. The ashes and decomposing vegetation (including underground organs) serve as fertilizer for the crop. Depending on the fertility of the native soil, roughly one to five successive crops can be planted before yields become low. The farmer then moves to another place in the forest and repeats the process. Native vegetation invades the abandoned plot.

Sedentary farming and herding made it necessary to measure crop area and farming time in calculating yields that would last through the winter. Mesolithic people were undoubtedly aware of seasonal changes and their effect on plants and animals. Neolithic agriculturists had to be more than aware of such change; they had to be able to predict when spring would come in order to ration their food supply and plan their planting. The need for calculation and measurement was to affect profoundly the subsequent development of human society and thought.

As groups became better and better at such calculation, farming supplemented by hunting and gathering gradually gave way to full-fledged agricultural economies. One effect of this development was to narrow the diet. Of the more than three thousand plant species used as food by different peoples throughout history, only about two hundred are domesticated, and of these fewer than twenty are cultivated intensively in any one area (see table 1). The reason is simple. Farmers learn how to cultivate efficiently no more than a few crops and then stick with them. Nonfarmers are likely to underestimate the knowledge required to cultivate plants. Not only must farmers know when to sow and when to reap, they must know how to prepare the soil and how deep to plant the seed. They must recognize pests that are likely to affect the stand and how to combat them. They must know how and when to harvest, and when it is not worth harvesting because yields are too low. They must be aware of water supply, and know when and how to irrigate, and how to manage the fallow, and how best to store the harvest. Furthermore, in field agriculture (as opposed to the garden agriculture practiced in moist tropical environments), it is difficult to plant many crops together because they have different requirements. Therefore farmers obtain better results by concentrating on a few crops well adapted to local conditions.[23]

Farming and herding, like so many human inventions, was a mixed blessing. Agriculture enlarged overall production at the possible cost of increased vulnerability of the food supply. Even with today's tech-

Table 1
Area, Yield, and Production of the
World's Twenty Most Productive Plants

Crop	Area (1,000 ha)	Yield (kg/ha)	Production (1,000 Tm)
Wheat	229,347	2,204	505,366
Maize	131,971	3,702	488,500
Rice	144,962	3,261	472,687
Potatoes	20,066	14,981	300,616
Barley	78,698	2,244	176,574
Manioc	14,010	9,676	135,551
Sugar (all kinds)	24,676	59,144*	121,524
Sweet potatoes	7,880	14,041	110,651
Sorghum and millets	91,859	1,139	104,592
Soybeans	52,683	1,914	100,809
Grapes	9,564	6,305	60,297
Oats	25,288	1,963	49,630
Cotton	34,712	—	49,823
Coconuts	—	—	41,040
Rye	16,738	1,989	32,288
Peanuts	18,728	1,106	20,708
Dry beans	25,665	581	14,909
Peas	8,832	1,494	13,199
Tobacco	4,111	1,595	6,559
Coffee	10,574	568	6,006

* Yield indicated is that of cane sugar.

nology, crops are susceptible to all manner of catastrophe—drought, disease, fire, war, infestation. Although there were calamities in pre-agricultural days, it is unlikely, for example, that a pest would wipe out all the wild plants eaten by a human band. In any event, the band could always move to another place. The great paradox of agriculture is that, while failing to eliminate the principal sources of mortality in hunter-gatherer societies, disease and accident, it introduced a new threat, famine and malnutrition. And because more people could be supported by an agricultural economy, an increased number were subject to these adversities.

Yet the development of agriculture in the Middle East, in the Far East, and in America ultimately led to the magnificent civilizations of Babylon and China and Mexico and Peru. Indeed, it was the foundation of all the great civilizations of this planet. Agriculture unleashed civilization, civilization unleashed human creativity, and human creativity bequeathed to us all the glorious art, and literature, and music, and philosophy, and science of which no other species is capable.

CHAPTER

Early Agriculture

IN EACH OF the five areas of the world where agriculture is known to have developed independently—the Middle East,
China, Southeast Asia, Mesoamerica, and the Andean Highlands of
South America—the process of domestication was slow. More than
2,000 years elapsed between the time some people began cultivating
edible plants and the time farming became their principal means of
subsistence. The Middle East, because it offers the earliest sites and the
best archaeological record, serves as a good model for the changes cultivation brought to plants and to the life of people. The oldest remains
of unmistakable agricultural villages lie in the mountains and hills of
the Fertile Crescent. Only a small number of the region's archaeologi-

cal sites have been excavated. Many are lowland localities associated with the monumental Mesopotamian and Egyptian cultures that flourished after 6300 BP, but numerous others, notably mountain sites, are older.

The region, an open hydrological basin, covers northwestern Iran, Iraq, southern Turkey, Syria, Lebanon, Israel, and western Jordan. It is roughly 950 kilometers wide from the coast of the Mediterranean Sea to the Zagros Mountains in western Iran, and some 1,050 kilometers long from the Anatolian Mountains in west-central Turkey to the Shatt-al-Arab, where the Tigris and Euphrates rivers discharge their silt-laden waters into the Persian Gulf. The region is an open hydrological basin that discharges its waters into the Persian Gulf. The mountains generally rise to over 3,000 meters, although in the southeastern region of Jordan they barely reach 1,000. The climate is Mediterranean, with winter rain and summer drought. Significantly more rain falls in the mountains than in the central basin, but rainfall differs from place to place and year to year. A number of rivers drain the mountains, and many of them flow into the interior basin. The most important of these are the Tigris and Euphrates, which originate in central Turkey and receive the waters of several tributaries before discharging them into the Persian Gulf.

During the Ice Age the region was dry and barren, with the typical steppe vegetation of low shrubs and grasses. Around 13,000 BP the climate grew wetter and warmer and, as evidenced by fossil pollen palynologists have recovered from the sediments of ancient lakes, the vegetation changed.[1] The mountains and hills of the Fertile Crescent turned into open oak-pine forest, the interior into a semidesert steppe. About 10,000 BP the climate stabilized and became similar to today's. At that time a dry forest dominated by oaks and pistachio trees grew in the mountain belt, with pines in the higher altitudes and a dry open steppe of shrubs and grasses in the interior. The mountain vegetation must have approximated that of today's central California coast ranges and the foothills of the Sierra Nevada. The vegetation in the dry interior was probably like that of certain deserts of Arizona and southern California.

Along with the transformation of plant life, the pattern of human

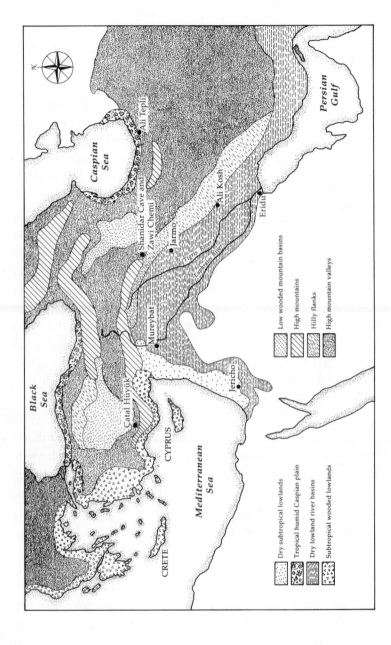

Early archaeological sites and vegetation types in the Fertile Crescent.

settlement changed. At the end of the Ice Age human populations in the Middle East were sparse and restricted to wetter areas, as shown by the remains of small camps of nomadic hunter-gatherers. When conditions improved, the number and size of camps increased and people advanced closer to the dry interior. They began to occupy permanent settlements between 11,000 and 10,000 years ago, probably before they developed full-time agriculture.[2] People settled permanently or semipermanently in the Levant (the westernmost zone, what is today Lebanon, Israel, and western Jordan), Anatolia, and the Zagreb-Mesopotamia area. For example, Jericho was occupied almost continuously between 11,000 and 8500 BP, Mureybat between 10,500 and 8500 BP, and Ali Kosh between 10,000 and 8000 BP.[3]

Several excavated sites indicate how agriculture began in the Middle East. An early site is Zawi Chemi Shanidar in Iraq, dated at 10,850 BP. It was probably a semipermanent camp with stone dwellings. Remnants of flint sickles and stone mortars reveal that the inhabitants ate seeds, probably cereal seeds. Most likely the people harvested the seeds but did not cultivate the plants. Along with bones of goats, remains of sheep, many of them lambs, were found at Zawi Chemi Shanidar. This represents a shift from the more usual pattern in the area of mostly goat remains and a balanced distribution of sizes. Some archaeologists have suggested that these animals were domesticated. The inhabitants, it seems, were starting to herd goats. They might have practiced some agriculture to supplement plant gathering, but they had not in any substantial way affected the wild plants they cultivated.

To the west, near the Mediterranean coast, there are several sites between 11,000 and 10,000 years old indicating the presence of Natufian people. Their tools—a large number of stone sickle blades as well as handles, mortars, pestles, and other grinding and pounding equipment—suggest that they ate cereals, which grow abundantly in the area. This and other indirect evidence suggests that wild wheat, barley, lentils, and other plants already served as important components of the diet of these pre-agricultural hunter-gatherers.

In Tell Mureybit, near the Euphrates River, archaeologists have found a large quantity of seeds of 10,000-year-old wild einkorn wheat and barley. Wild einkorn wheat does not grow there today; the nearest site is some 150 kilometers away. Perhaps it grew over a larger area in the past, or perhaps the people at Tell Mureybit traded for it. Another explanation is that they were starting to cultivate wheat there.

The first concrete evidence of plant cultivation is from Tell Aswad in Syria. People grew here domesticated peas, lentils, emmer wheat, and barley in the earliest phase of the settlement, dated at 9800 BP. Another site where there is unmistakable evidence of early agriculture is Tepe Ali Kosh in Iran. In beds dating from 9500 BP, archaeologists found abundant plant remains, including the seeds of many wild pulse plants, cultivated emmer wheat, wild and cultivated einkorn wheat, and barley. There are also clear indications from fossil bones that the people here had already domesticated both sheep and goats. From a site at Jarmo in Iraq, dated at 8750 BP, there is unmistakable confirmation of cultivated two-rowed barley and emmer and einkorn wheat, and of the existence of domesticated goats, sheep, and pigs.[4] Remains found in many nearby sites show that by 8000 BP people in this area grew wheat, barley, peas, and lentils and kept sheep and goats.

Barley, lentils, and wild einkorn and emmer wheat were the basis of early agriculture in the Middle East. Wild wheat and barley still grow throughout much of the Fertile Crescent. Agronomists calculate that in a rainy year, 500 to 800 kilograms of grain per hectare are produced naturally in stands of wild wheat and barley on the east Galilee uplands.[5] The ethnobotanist Jack Harlan, employing stone tools similar to those used by people 10,000 years ago, in one hour collected 1 kilogram of clean wild wheat seeds in the mountains of southern Turkey.[6] He calculated that at that rate, in a three-and-a-half-hour day a person could gather enough wild wheat seeds to last for eleven days.

People in the Middle East domesticated a number of the plants that to this day are commercially important. These species were probably already an important element of the local economy. Once domesticated, the crops were quickly adopted by neighboring people in similar stages of cultural development.

Barley and wheat are the oldest cultivated cereals. Barley is native to the Fertile Crescent. Wild barley often grows in the company of wild wheat. It withstands drier conditions and poorer soils than wheat and will also grow on soils that are mildly saline. In marginal areas, barley

Spikelet and spike of two- and six-rowed barley, the first cultivated cereal.

yields more grain than wheat. Wild barley, what we call two-rowed barley, has three florets per spikelet, of which only the middle one is fertile. All early archaeological remains of barley are two rowed. During domestication farmers produced six-rowed varieties where all three florets in a spikelet were fertile. Another change under domestication was the production of naked seeds, that is, seeds that lost their hulls on threshing.

Barley, eaten in porridge or in some form of flat bread, was an important component of the human diet up to the end of the Middle Ages. Because barley flour has no gluten people mixed it with wheat flour to make a low-quality "poor man's bread." Today we use barley almost exclusively as animal fodder and for brewing beer. Different varieties are used for these two purposes. To make beer, the seeds have to first be germinated. This converts the starch into sugar, which the yeast then transforms into alcohol. This process, called malting, was discovered early on. The ancient Babylonians and Egyptians made beer regularly. Today barley is the fourth most common cereal.

Wheat is the most abundant plant on earth. Approximately 15 percent of the arable surface of the world, some 229 million hectares, is planted with wheat. It accounts for more than 20 percent of all calories eaten by people. Wheat is nutritionally superior to the other main cereals. Its seeds have not only starch (60 to 80 percent of dry weight) but also a significant quantity of protein (8 to 14 percent) and adequate amounts of most essential amino acids, fats, minerals, and vitamins such as E and the B complex. Gluten allows it to rise when leavened, so that on baking it makes bread, an appetizing, nutritious, and easily digestible product. Wheat is a staple food for about two billion people, about a third of the world's population. Wheat is the traditional food staple of Europeans, who took it to Australia and the Americas. It is also the principal food staple in the former Soviet Union, the Middle East, northern India, and northern China. Today farmers cultivate wheat in cold- (Canada, the former Soviet Union) and warm-temperate regions (the United States, Australia) and in subtropical areas (Brazil, India).

Wheat is not one species but a complex of species, some wild and others known only from cultivation. It was first domesticated about 10,000 to 9,000 years ago. The first cultivated plants were einkorn wheat (*Triticum monococcum*). Wild einkorn (*T. boeticum*) grows in the Fertile Crescent, western Asia, and the southern Balkans, on dry hillsides, where it can be quite abundant. It also grows as a weed in disturbed habitats such as the edges of fields and roadsides. Cultivated einkorn differs from its wild cousin in having a less fragile spike, which at maturity does not shed its seeds. It is a low-yield species, cultivated today only in isolated pockets of Asia Minor and the Swiss Alps. Einkorn is a small plant, but it can grow in poor soil unlike other wheat species. It makes a yellow flour, poor for baking. People consume einkorn primarily as porridge.

Another wheat species known in both wild and cultivated forms is emmer. Wild emmer (*T. dicoccoides*) grows throughout the Middle East. Cultivated emmer (*T. turgidum*, subspecies *dicoccum*) is seldom grown today but was the main domesticated wheat of the Mesopotamian cultures and early European agricultural settlements. Both emmer and einkorn have covered or "hulled" seeds, which are characteristic of wild grasses. The hulls protect the seed from predators and aid in dissemination.

Spikes of wheat: einkorn (left); emmer (middle); common, or bread, wheat (right).

Emmer and einkorn differ in some respects, most importantly in the number of chromosomes. Chromosomes contain the DNA, the chemical substance responsible for the inherited characteristics of organisms. Einkorn wheat has seven pairs of chromosomes and emmer has fourteen. Seven of emmer's chromosome pairs are identical with einkorn's; the other seven presumably come from a wild grass that grows in the area where einkorn is native, one of the so-called goat grasses.[7] Sometime in the past, einkorn wheat crossed with goat grass, producing a hybrid with fourteen chromosomes, seven from each parent. Hybrids such as this, which lack identical pairs of chromosomes, tend to be sterile. Occasionally in a cell of a sterile hybrid, however, there is duplication of chromosomes. This process restores fertility and accounts for the origin of fertile emmer wheats. Plants with two of each kind of chromosome are called diploid. Those where the normal diploid number has been doubled, such as emmer, are tetraploid.

There are other cultivated wheat species with fourteen pairs of chromosomes, among which the most important is durum wheat. It differs from emmer in having naked seeds that thresh free. This is of course a great advantage when it comes to milling, but it reduces the species's ability to grow in the wild. Farmers cultivate durum wheat for the "harder" quality of its flour (semolina), which makes it ideal for making couscous and pasta.

Some 80 percent of the wheats grown today have not fourteen but twenty-one pairs of chromosomes (hexaploid). Presumably another species of goat grass provided the third set of chromosomes. These are the bread wheats, of which one species, *T. aestivum*, is the most widely cultivated, accounting for over 70 percent of the world's wheat production. Bread wheats, which also have naked seeds, are only known in cultivation; they might have originated spontaneously in cultivated fields 4,000 or 5,000 years ago. Farmers and agronomists have profoundly modified them by breeding, and today there are more than 17,000 varieties.[8]

Cultivated oats are a nutritious cereal, containing about 16 percent protein and 8 percent fat. Native to the Near East and the Mediterranean basin, they grow well in moist climates in temperate latitudes, where they often outperform wheat. They probably became weeds in grain fields and were later cultivated in their own right. People originally ate oats in some form similar to the breakfast porridge still eaten today, yet their major use was and continues to be as a fodder plant, especially for feeding horses.

We use rye as an early-spring forage and to make bread and rye whiskey. It was more important in the past than today. Rye, the only cereal other than wheat that has gluten in its flour, will make a leavened bread. But it does not have as much gluten as wheat and its flour is not as elastic. To improve the quality of rye bread, bakers mix rye flour with wheat flour. Rye has another important characteristic: It will grow in

Panicle of spreading oats.

cold climates and mature even if summer temperatures do not rise above 15° C (59° F). It is therefore common in northern Europe and northern Asia. Rye is also resistant to drought and can grow in sandy or acid soils. Thus it succeeds under conditions where wheat often fails. Originally people cultivated rye all over Europe, but after the fifth century A.D. they grew it mainly in the northern countries, especially the Baltic. It commanded a lower price than wheat, its yields being slightly higher and more dependable.[9] A peculiar characteristic of rye is that it has to be crosspollinated (a characteristic it shares with maize). Yields of rye fields are dependent on effective crosspollination by means of wind.

Rye, like oats, probably originated as a weed in wheat fields. It is native to the Anatolian plateau in Asia Minor, where farmers cultivated it in early Neolithic times. [10] Most likely it was one of the many grass species whose seeds hunter-gatherers collected in that area. When it appeared in wheat fields farmers would have continued to harvest it, not considering it as a weed in the modern sense. It probably moved into Europe with early wheat agriculture. Eventually farmers, recognizing rye's hardiness, started to grow it as a separate crop. The first definite evidence of rye cultivation in Europe comes from Bronze Age settlements (1800–1500 B.C.) in Czechoslovakia.

Lentils are among the oldest—if not the oldest—domesticated legumes, having been grown together with wheat and barley since the beginning of agriculture 10,000 years ago. Archaeologists have found seeds of wild relatives of the lentil in pre-agricultural settlements at Mureybit and Tell Abu Hureyra in Syria together with wild wheat and barley. From many ancient farming villages in the Middle East they have recovered carbonized seeds dated 9000 to 7000 BP. Lentils were also found among the remains of Neolithic farming communities in ancient Egypt and of the first farming villages in Crete and the Balkans.

The lentil is a slender, tufted, multibranched annual plant with small lens-shaped seeds. Yields are low compared with cereals. Lentils are a tasty and nutritious food containing about 25 percent protein by weight. People eat lentils in soup or stew or mixed with wheat or rye. The lentil was an important component of the diet in ancient Greece and Rome. Today the lentil is cultivated widely in the Mediterranean area, northern Africa, Ethiopia, the Near East, and east to Pakistan and India. It has been introduced into the Americas, where the principal producers are Argentina, Chile, and the United States.

The common pea is another Middle East crop that has probably been under cultivation for as long as lentils. Peas are annual, climbing, or trailing plants with white or colored flowers. The seeds are highly nutritious, about 22 percent protein. Peas were an important crop in the Mediterranean in classical times and in Europe in the Middle Ages. Today they are the second most important seed legume crop in the world; more than ten million tons are produced annually. Peas are grown primarily in temperate areas, especially northern Europe, parts of Russia, China, Canada, and the United States. In the past primarily a garden crop, recently they became an important field crop for the frozen-food industry in Europe, the United States, and Canada. While in the past peas were cultivated for their mature dry seeds, today farmers harvest them green as "fresh" peas. Botanists do not know either the wild progenitor or the early history of the pea.

The broad or fava bean, about a quarter protein, is an erect plant characterized by its easily threshed pods and large seeds. People eat it in both dry and green form. Also known as the horse bean, it is a good forage plant. By the Iron Age the culture of the broad bean was firmly established in Europe and Egypt. It was introduced into China some 700 to 800 years ago and has become an important crop there. Today China is the largest grower. Other important producers are Italy, Spain, northern Africa, Mexico, Peru, and Brazil.

The chickpea originated in western Asia, probably in the Caucasus area. It is a branching, bushy annual adapted to warm climates. The earliest archaeological remains of chickpeas come from a site in Turkey dated at 5450 B.C. From there it diffused both westward into the Mediterranean and eastward to India, where it arrived some 4,000 years ago. Chickpeas are an important food staple grown in India (the primary producer), the Mediterranean basin, Ethiopia, and Mexico. They are highly nutritious, with 20 percent crude protein, 80 percent digestibility, 50 to 60 percent carbohydrate, and about 5 percent oil. A diet of equal portions of chickpeas and cereals meets human amino acid needs. People eat chickpeas as dry seeds in a variety of ways, and they are an important meat substitute in traditional peasant communities. In India people often process chickpeas into dhal (split pea) or flour, but they also consume them whole, boiled, or roasted. The chickpea is the third most important seed legume, with some 7 million metric tons of annual production.

It took at least a thousand years, and possibly more, for agriculture to become the principal means of subsistence for most of the population of the Middle East. One place where archaeologists have carefully documented the transition from hunting-gathering to agriculture is Jericho.

Jericho, located slightly north of the Dead Sea, was one of the first permanent settlements in the Middle East. The earliest settlement was not much more than a collection of houses scattered over an area of about four city blocks. It is remarkable only for the existence of a tower built of stone and mud, which is the first example anywhere of "monumental" architecture, though considering later developments it is a very modest structure.

Archaeological evidence suggests that the first inhabitants of Jericho were learning to cultivate plants and domesticate animals. The later expansion of Jericho probably resulted from rapid improvement in agriculture and animal husbandry in this fertile oasis. It is possible that people practiced some primitive irrigation at a slightly later stage. Population probably expanded with the food supply. Defenses involving expansion of the tower and construction of a protective wall around the settlement date from approximately 10,000 years ago. Presumably erected to protect stores of food, they indicate that the larger population could not have lived solely from hunting and fishing or even primitive agriculture. Trade must have played a part in the town's development. Jericho commanded supplies of salt, bitumen, and asphalt from the Dead Sea. Eventually, unknown invaders destroyed the city.

Another early permanent settlement was Catal Hüyük in western Anatolia.[11] This was much larger than Jericho, some thirteen hectares in extent. It lies sixty-five miles north of the Mediterranean Sea in the plain of Konya (Iconium), now a salinized steppe, then a grassland with oaks and pistachios. People lived there from about 9,500 to 8,500 years ago. This site was especially lush for the region, and the community may have achieved a degree of control over a wide section of the Konya Plain. Catal Hüyük was probably an important trading center. Here we find exquisite mural paintings, evidence of craft specialization, and more permanent and better-built dwellings than at Jericho. The earliest occupants of Catal Hüyük seem to have had domesticated sheep and goats. Archaeologists have found remains of fourteen cultivated plants, but emmer wheat, barley, and lentils were the staple foods, obtained almost entirely from cultivation. Carbonized fragments of woven and twined material, probably flax, have been recovered, as well as the earliest-known warp-weighted loom.

The Zagros highland in western Iran and Iraqi Kurdistan is an area with many archaeological sites indicating the emergence of settled, food-producing communities in the Middle East. The earliest archaeological sites date from between 10,000 and 9000 BP. By 8000 BP some locations had evolved into permanent agricultural villages. The archaeological and botanical evidence strongly suggests that during this time people shifted from primarily hunting and gathering to primarily growing plants and domesticating animals. The same story applies to what is today western Iran and the area bordering the southern rim of the Caucasus and the Caspian Sea, where people domesticated the olive and the wine grape some 6,000 to 5,000 years ago.

These early villages are all in geographic areas that had a mixture of vegetation to support a variety of plants and game animals. Hunting and gathering must still have played an important role in the economy. Given all the uncertainties of early agriculture, people would have been reluctant to break completely with their traditional way of getting food. The diversity of habitats allowed them to try out early techniques of planting and cultivating while still living from hunting and gathering.

Studies conducted at these farming sites in the hills help explain the origins of agriculture. First, all indications are that the populations had enough food throughout the period farming was being adopted. It does not appear that people turned to planting because they were hungry. Second, populations were probably high for a hunter-gatherer economy. (Figures on the size of prehistoric populations are difficult to estimate.) Third, the concentration of individual households indicates that people had become sedentary. Finally, from artifacts found in graves there is some evidence of a disparity in wealth between the inhabitants of a given village. In other words, society was growing more hierarchical.

The hilly areas of the Middle East seem to have supported the transition from hunting and gathering to planting, but they were not necessarily the lands best suited for agriculture. For best results plants must be grown on flat land with rich soil. In the hills people came up with the idea of agriculture and developed the basic techniques. In time, others would take agriculture to the river plains and adapt it to the new terrain there.

Once crops became domesticated in the Middle East, they and the art of planting diffused to neighboring peoples who were in similar stages of cultural development. Middle East agriculture slowly spread into central Asia. Archaeologists have found remains of a barley-wheat-peas-and-lentil agriculture in the Kashmir area of India by around 4500 B.C.[12] In time, Indian farmers domesticated many local plant species. Wheat and barley cultivation reached China at about the same time it reached India. There is, however, indisputable evidence for an independent prior adoption of agriculture in China and Southeast Asia based on millet and rice cultivation.

North China includes the Huang Ho or Yellow River basin surrounded by the Mongolian plateau to the north and the Tsingling Mountains to the south. In the northwest is a large, high plateau formed by a thick layer of fertile loess, a fine-grained soil deposited by the wind. Central China includes the vast and fertile basin of the Yangtze River. South China encompasses the plateau of Yunnan and Kweichow and the valleys of the Si and Canton rivers. To the west China borders the highlands of Tibet, and to the south it borders the Indochinese peninsula and Burma. China is a vast area of more than 3.5 million square kilometers. The climate ranges from the cold temperate north to the subtropical south. Precipitation increases from north to south and is greater in summer than winter.

Of the five major areas of the world where farming was adopted independently, we know the least about China. The oldest evidence of agriculture there comes from the area of the Wei River valley in Shenshi Province.[13] This is really the center of the country, but we refer to it as the northwest because it is so in relation to the ancient core of China. The area is mountainous, about 500 meters above sea level, dry, with hot summers and cold winters, and ecologically diverse, with pockets that are much wetter than the area as a whole.

Around 8000 BP the Neolithic Yang-shao people were cultivating

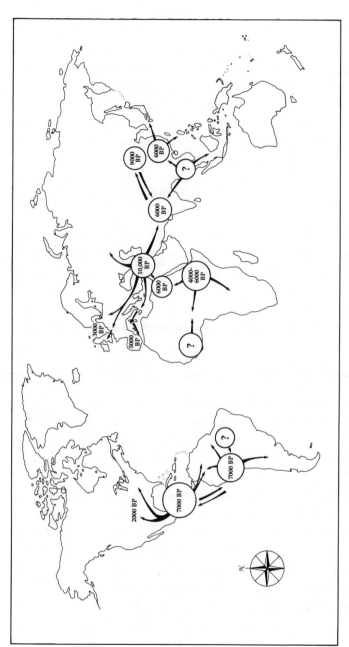

Areas (circled) where agriculture originated, with approximate dates and the general direction of its spread. Circles with question marks indicate the possibility of independent origin but no proof as yet.

foxtail, or Italian, millet and panic millet and had domestic pigs, dogs, and chickens.[14] Archaeologists have found 8,000-year-old remains of foxtail millet and common millet in farming communities in north-central China. By 6500 BP an agricultural system based on a variety of plants had developed. Besides the millets, there were Chinese cabbage, walnuts, pine seeds, chestnuts, and bamboo shoots. Some plants, such as bamboo shoots, persimmons, grass seeds, walnuts, pine nuts, and chestnuts, may have been gathered from the wild rather than cultivated. Storage jars are common in the early archaeological sites. There is also abundant evidence that the Chinese cultivated mulberries and had domesticated silkworms.[15] They produced pottery and a diversity of Neolithic stone tools.

Millet is not one but a variety of grass species with small seeds used as forage or human food. Today foxtail millet is a minor crop in southeast Europe, several parts of Asia, and North Africa, a more important one in India, China, and Japan. Other millets are Japanese, or sanwa, millet, common in Japan in areas where rice will not grow; bulrush, or pearl, millet, a major cereal in tropical savannas of Africa and India; and ragi, cultivated in Indonesia and Africa, which gives high yields even in poor soil. It is probable that Chinese and Mediterranean farmers domesticated common millet independently.

The soybean was among the plants domesticated in northern China, also its center of origin, some 4,000 years ago. This small, erect or prostrate annual plant is today the most important legume in the world in terms of production, area under cultivation, and international trade. Soybean seeds contain 35 to 45 percent protein and 20 to 23 percent oil. In the Far East, people use soybeans for oil or powder or in curd form as human food. Immature green beans and sprouts are highly nutritious and the Chinese eat them in great quantities. In the West, we consume soybeans as oil or meal. The oil is converted into margarine, shortening, mayonnaise, and salad oil. Most of the meal is added as a source of protein to feed for poultry and hogs. It is also used in the preparation of meat substitutes. Until the twentieth century soybeans were grown almost exclusively in Asia. They did not become an important crop in the United States until the 1930s. Today the United States is the world's greatest producer (and consumer) of soybeans, followed by China, Brazil, and Argentina.

The soybean plant.

Rice, the staple crop most commonly associated with China, and wheat, an equally important crop in northern China where rice cannot grow, were later additions to the diet. Wheat was imported from the Middle East by way of India. Rice was probably domesticated in Southeast Asia and brought to China. There are 7,000-year-old remains of rice excavated from different parts of China but primarily from Zhejiang, which many believe to have been the original area of rice culture in China. Five-thousand-year-old Chinese writings mention rice as one of the five most important food plants. Rice was cultivated then, as now, in south China, where the human population was greatest. In the early days, rice culture in China consisted of broadcasting seeds in natural marshes without plowing. Irrigation was a later development (approximately 2,500 years ago), as was transplanting.[16] We know far less about the domestication of rice than of wheat or maize. Rice was probably first domesticated in Southeast Asia, in what today is Thailand, between 8,000 and 7,000 years ago. Most likely from there it was taken east to China and west to India, where it became the mainstay of their civilizations. Or it might have been domesticated in all three places at about the same time. [17]

Although some historians have claimed that agriculture was independently developed in northern India, there is no good supporting evidence. A civilization developed in the valley of the Indus River around 3000 B.C. around the substantial cities of Mohenjodaro and Harappa. Yet the evidence uncovered in these sites shows that Indian farmers cultivated mainly wheat and barley, which come from the Middle East. India did, however, play an important role in the domestication of rice and legumes. Rice is one of the most diverse crop species. Some fifty thousand varieties are known, of which twenty thousand come from India. Archaeologists have found rice grains about 8,000 years old together with the remains of a Neolithic culture at Maharaga, Uttar Pradesh. Excavations at Mohenjodaro, near the Pakistan border, uncovered rice grains in an earthenware vessel that was approximately 4,500 years old. Impressions of paddy rice and husks were also discovered on 4,500-year-old clay tiles in Lothal. Carbonized paddy grains and husks have been found in archaeological sites in Hastinapur, Uttar Pradesh, that archaeologists estimate to be about 3,000 year old. The spread of rice cultivation to central and eastern India seems to have occurred about 3,500 years ago. The oldest sacred book in Sanskrit, the *Rig-Veda*, does not mention rice, although it mentions wheat and barley, introduced from the Middle East probably by the Aryans. The *Atharva-Veda*, from about 2000 B.C., does mention rice.

Although throughout the world less arable land is dedicated to rice than to wheat, rice feeds more people than any other crop. Rice is nutritious although relatively low in protein (7 percent). It is the principal staple food of the humid tropics and subtropics, an important food of many people in the Americas as well as the basis of oriental cuisine. Rice flour does not have gluten and therefore cannot be made into leavened bread. People usually eat rice boiled or in porridge.

There are about twenty-five wild species of rice, but only one (*Oryza sativa*) is widely cultivated. What is called wild rice in the United States is not truly rice but a wild grass that grows in marshes and ponds. It was and still is harvested but not cultivated by American Indians. Today wild rice is cultivated in California.

Panicle of rice.

Rice diffused to all of Asia in prehistoric times. There are indications of its cultivation in Indonesia and the Philippines 4,000 years ago and in Japan and Sri Lanka 3,000 years ago. Europeans did not cultivate rice during classical times, but the Romans knew about it. The Greek historian Diodore of Sicily, who lived in the first century A.D., described the plant and its cultivation from reports obtained by Aristobule, who took part in the expedition of Alexander the Great to India. The Arabs introduced rice cultivation into the Mediterranean together with irrigation techniques. The Spaniards and Portuguese took rice to the New World. It was brought to the United States early in the seventeenth century.

An important pulse crop in India, its presumed native site, is the pigeon pea, grown mostly by small subsistence farmers. The seeds are processed into dhal, or dried split pea, which stores well. The pigeon pea is a tall shrub and a promising forage plant. The exact place of its domestication is unknown but most likely India. There are Sanskrit references to the pigeon pea going back 1,600 years.

Humans are relative latecomers to America. They crossed the Bering Sea from Asia during the last Ice Age, somewhere between 70,000 and 30,000 years ago. From Alaska people slowly spread south until they reached the tip of Tierra del Fuego at least 10,000 years ago.[18] They adopted agriculture in the New World a little later than in the Old World, perhaps by as much as 2,000 to 3,000 years. Planting developed independently in Mesoamerica (central and southern Mexico and Guatemala) and in Peru.

Although most plant species brought into cultivation are unmistakably native to one or the other region, some species such as maize and beans were cultivated in both, indicating that perhaps exchanges of crops and techniques took place. This overlap has generated a fair amount of controversy, raising the question of whether agriculture really arose independently in two distinct parts of the New World. Those that favor two independent origins point out that farmers in Mesoamerica grew mostly plants that reproduced by seed (maize, beans, squashes), while many of the plants grown by farmers in Peru are root crops (manioc, potatoes, ulluco, oca).[19] An alternative hypothesis is that agriculture started in tropical South America and from there diffused both north and southwest. Once given the idea of agriculture, people in Mesoamerica or Peru could have applied it to domesticate local plants.

Mesoamerica is a high plateau bordered by two parallel mountain chains, the Sierra Madre Occidental and the Sierra Madre Oriental, flanked by narrow strips of tropical vegetation between the mountains and the sea. The climate is tropical monsoon, with summer rain and winter drought. Monthly average temperatures vary according to altitude, from 75 to 85°F (25 to 30°C) at the coast to 65 to 75°F (20 to 25°C) in the central plateau, and are even all year round. Precipitation is adequate to maintain a rich tropical forest in the lowlands and mesic pine forests at higher altitudes. However, there are arid pockets created by the mountains, called rain shadows.

People probably arrived in Mesoamerica about 20,000 years ago, although the oldest human remains, discovered near Tepexpan, a site twenty-five miles northeast of Mexico City, are only 9,000 years old.[20] The inhabitants of the region developed an elaborate hunting and gathering economy using such plants as century plant, prickly pear, mesquite, and oak acorns. By 7000 BP Mesoamerican farmers were growing maize (corn), squash, chili pepper, avocado, and amaranth in southern and central Mexico.[21] They also domesticated common beans, tomatoes, cacao, papayas, guavas, vanilla, cotton, and sisal for fiber.

The best evidence for the origins of agriculture in Mesoamerica comes from the Tehuacán Valley in south-central Mexico. This is a dry valley surrounded by hills. People moved into the area around 12,000 years ago and occupied the caves in the hills. They have left behind one of the best records known of the early evolution of a culture and the adoption of agriculture. No one knows whether people here learned to plant crops independently or from people living in other areas of Central America. Archaeologists have also recovered remains of cultivated plants from caves in Tamaulipas to the north and Oaxaca to the south, but these remains are not as complete or as rich as those found in the Tehuacán Valley.

The Tehuacán people, like others in Mesoamerica, depended for many millennia on hunting and gathering. Then some 9,000 years ago they began to cultivate some of the plants in their diet: maize, squash, chili peppers, avocado, and amaranth, and a little later beans. Archaeologists have found remains of cultivated plants in the caves that line the valley. They have also found the remains of dogs, known to have been eaten in ancient Mexico. The continuous archaeological record of the Tehuacán Valley tells us that the early agriculturists were seasonally migrating people who became sedentary long after they had domesticated maize.

Maize is the most important botanical legacy of Mesoamerica. The Tehuacán Valley has the oldest of many archaeological sites containing remains of maize. However, the site has not yielded remains of teosinte, the presumed wild ancestor of maize, nor does that species grow there today. The earliest cobs found in Tehuacán, although small and primitive by all accounts, are considered cultivated forms by most botanists. This has led to the hypothesis that the inhabitants of the Tehuacán Valley acquired the crop rather than domesticating it.[22]

Once the people of central Mexico started cultivating plants they increased in numbers and became sedentary. The lack of large domesticated animals meant that the methods of cultivation would be different from those in the Middle East. Hoe and planting stick were the principal agricultural implements, and mixed cropping prevailed. A common practice was to plant maize, beans, and squash together, a method still employed in traditional agriculture in the Americas from Mexico to northern Argentina. Beans, being legumes, helped fertilize the field. As everywhere else, early agriculture was of the swidden type. The Lancandon Maya and other indigenous groups in Mexico and Guatemala still make clearings in the forest for their *milpas*, or fields, where maize and other mixed crops are planted.

Maize produces the greatest bulk of seed per acre of all cereals on earth. It is rich and nutritious human food but is used in many developed countries, such as the United States, as animal feed. Maize is the principal staple for many people in Central and South America as well as the principal cereal eaten by many African populations. It can be made into flat, unleavened bread, such as tortillas in Mexico or *arepas* in Venezuela. People also eat maize as a porridge or gruel (polenta, *mazamorra*), or boil or steam the grains (corn on the cob).

Maize is unusual as a grass in bearing its male and female flowers separately. Its plants bring forth the male flowers terminally, in spikes called tassels, and the female flowers halfway down the stem, in lateral structures called cobs or ears. Maize is also distinctive in that its seeds are not covered by floral bracts (glumes, lemma, palea). Instead, the seeds are exposed on a massive axis, the cob, the entire structure being enclosed and protected by leaf sheaths.

Over the years botanists have argued ardently about the ancestry of cultivated maize. The two principal theories are that maize descends from a species of the related genus *Teosinte*, or that both maize and teosinte have a common ancestor.[23] Although the arrangement of the flowers of teosinte is very different from that of modern maize, one species, *Zea (Teosinte) mexicana*, and maize hybridize regularly in central Mexico, where teosinte is a common weed in maize fields.[24] Many botanists are beginning to accept the view that maize is the cultivated form of a species of teosinte.

The cob of modern maize is unlike anything found in the wild. Early maize cobs found in Tehuacán are different and much more like teosinte. Maize is an excellent example of how plants are modified by cultivation. Instead of a giant cob, early maize had a thin spikelet that pro-

An ear of modern flint corn.

duced female flowers on the lower part and male flowers on the upper part. This structure was formed erect at the top of the plant. Furthermore, the cobs had long protecting glumes that partly or totally enclosed the seeds, as in most other grasses. At maturity the seeds probably broke loose from the cob and dispersed individually. These are the characteristics of modern teosinte, and the reason it is believed that the massive cob of modern maize is the result of human selection after teosinte was domesticated.

Remains confirm this interpretation. Maize cobs from archaeological sites in Mexico and the southwestern United States are larger and more varied in their morphological characteristics than the oldest cobs found in Tehuacán. The larger cobs are presumably the result of human selection and better growing conditions. The cobs are also harder and thicker, the glumes do not enclose the kernels, and the leaf sheaths are larger.

Maize was taken from Mexico to South America in pre-Columbian times. In Peru, where it was introduced at least 5,500 years ago, farmers

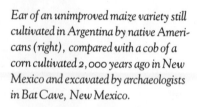

Ear of an unimproved maize variety still cultivated in Argentina by native Americans (right), compared with a cob of a corn cultivated 2,000 years ago in New Mexico and excavated by archaeologists in Bat Cave, New Mexico.

produced many races, including the one with the largest seeds and the one with the longest cob. After the discovery of America, the Spaniards took maize to Europe, where it became an important staple in southern Europe, especially Italy. The Portuguese took maize to Africa.

Beans were domesticated early on in both Mexico and Peru. Highly valued all over the world, they are consumed in the tropics mostly in dry form but as both dry seeds or fresh pods (string beans) in temperate regions. Low-fiber varieties have been developed for consumption as fresh green beans and for the frozen-food industry. Beans are an important source of protein, and with maize or rice form a nutritious and complete diet. They are low, erect, or twining annuals with trifoliate leaves and white or pink flowers. There are four species of domesticated American beans. The common, or kidney, bean was apparently domesticated independently in Mexico and Peru. Anthropologists have dated kidney bean remains from the Callejón de Huaylas in Peru at 7680 BP, and from the Tehuacán Valley in Mexico at 7000 BP. The lima bean was also domesticated in both North and South America. Forty-five-hundred-year-old remains of large-seeded lima beans have been found in Huaca Prieta, Peru, small-seeded 1,800-year-old limas in Mexico. Tepari beans were domesticated in Mexico at least 5,000 years ago, while the oldest runner bean remains are 2,200-year-old fragments from the Tehuacán Valley.[25]

Beans under cultivation have lost their forerunners' ability to disperse seeds at maturity. Wild beans are slender trailing vines, while cultivated beans are either short twiners or small shrubs. There has also been selection for larger pod and seed size. Farmers cultivate all four species of beans as annuals, but in the wild both tepari beans and lima beans are perennial. Traditionally beans are cultivated in the tropics for their seeds, which are eaten as dry beans.

Peru and the highlands of Bolivia, like the Middle East and the Tehuacán Valley, are not areas where one would expect independent development of agriculture and plant domestication. This is mostly dry, barren land surrounded by high mountains. The coast, particularly in southern Peru, is the driest in the world. Humans can survive only in the moist river valleys that bring water from the mountains.

Further inland there are two parallel chains of high mountains separated by a series of longitudinal valleys. These mountain chains have been called since Spanish colonial times the Sierra. The Incas called them the Keshua. The vegetation at the bottom of the valleys is dry scrubland, but on the flanks of the mountains and along the streams there is forest vegetation. The easternmost chain of mountains splits into two chains at latitude 8° S and then reunites at roughly latitude 30° S.

Between the two chains extends a high, flat plateau known by its Inca name of Puna. It ranges between 3,500 meters in the north to 4,000 meters in the south. Here the climate is harsh. In the upper reaches frost and snow occur throughout the year. Soft grasses and sedges grow along the Puna's water courses and in wetter areas called mallines, and hard, bunchlike grasses grow in exposed places. The Puna shows a humidity gradient decreasing from north to south, where it is very dry. Lower elevations are both more moist and slightly warmer. Roughly west of center lies Lake Titicaca, the highest lake in the world. Around the lake several civilizations developed, of which Tiahuanaco and Quechua (the Inca civilization) are the most important. Although situated in the geographical tropics, for the most part this area too has a climate made harsh by drought and altitude.

The oldest human remains from Peru are dated at 14,000 BP, but there are tools and other human artifacts that suggest the presence of humans in Peru for at least 15,000 years.[26] These early people were nomadic hunter-gatherers who specialized in the hunting of large game, especially the now-extinct giant sloth. Some bands lived near the coast and roamed from there to the flanks of the Sierra. Others lived in the Sierra and wandered up to the Puna in search of vicuñas and guanacos, relatives of Asiatic camels.

About 12,000 BP the climate of the region, which had been very cold and dry, grew wetter and warmer. The humid forestland flanking the mountains expanded, and with it apparently also the population. Remains of human camps from these times are more common than from earlier times, and the tools more numerous and elaborate. Seed-grinding implements make their appearance, indicating that small, hard seeds such as grasses or leguminous seeds were becoming an important component of the diet. There is no indication of plant cultivation at this time.

Some 9,000 years ago the population increased, and people in the mountains started to exchange goods with those on the coast. Signs appear of incipient plant cultivation, but not of a sedentary life-style. By 8500 BP people of the dry western Sierra at the site of Guitarrero Cave were starting to grow chili peppers and beans.[27] In the Ayacucho Valley archaeologists have found evidence of the cultivation of quinoa and perhaps also squash between 7800 and 6600 BP.[28] There are also indications of the domestication of quinoa in the Junin Valley around 7000 BP.[29] Quinoa is a pseudo-cereal, that is, it has small, hard seeds but is not a grass. It might have served the role that true cereals played in other areas where agriculture arose. Early evidence of the domestication of other plant species comes from Huaca Prieta in northwestern coastal Peru, where 5,000 years ago people cultivated gourds, squashes, cotton, lima beans, and chili peppers.

The picture that emerges is of the arrival 15,000 years ago of bands of hunter-gatherers, presumably from the north. They slowly spread throughout this area, some settling in the fertile valleys of the coast, which offered its rich sea life, others in the high valleys of the Sierra, and still others in the high Puna or in the subtropical and tropical eastern flanks of the Andes. These bands were nomadic and small at the beginning, coming from a diverse range of ecological zones. As time went by they grew larger and more sedentary, and some 9,000 to 5,000 years ago they started planting and domesticating some of the plants they were eating. The situation appears different from that in the Levant, where sedentariness in the Natufian people seems to have preceded or developed along with agriculture. Apparently in South America nomadic people gradually added cultivated plants to gathered plants and game. At about the time that people began farming, they domesticated the guinea pig and the guanaco. They kept guinea pigs in pens for eating, as Andean Indians still do today. The guanaco, bred primarily as a beast of burden, evolved into the llama, which also yields a coarse wool. Bred for its wool, the guanaco evolved into the alpaca, which produces more and better quality wool than the llama but is a mediocre beast of burden.

After 5000 BP there occurred a period of village-centered agriculture similar to that in the eastern hemisphere and Mesoamerica. This was the predominant social organization for 2,000 to 3,000 years. The first archaeological remains of cities in South America are those of Chavin in north-central Peru, approximately 3,000 years old.

Living in such a heterogeneous area, the inhabitants domesticated plants from three distinct ecological zones: first, the moist rainforest on the eastern flank of the Andes, where the ancestors of manioc, peanuts, guava, coca, and lima beans grow; second, the mountain plateau, home of the potato, quinoa, oca, and ulluco; and third, the coastal desert, home of cotton. Most likely different bands cultivated different species and then passed on their knowledge in exchanges with other villages.

The kidney bean and the peanut, or groundnut, are important legumes native to South America. The peanut is a herbaceous low-lying herb that buries its fruit below ground once its flowers have been pollinated. Farmers grow it primarily for oil, which constitutes roughly half of the seed, but it is also rich in protein (25 to 28 percent) and has a good amino acid composition.[30] Peanuts are an important component of the human diet in parts of South and North America, Africa, and Asia. Grown in tropical and subtropical countries around the world, mainly in India, Africa, and China, they prefer loose or sandy soils.

The cultivated peanut is a tetraploid (a plant with twice the normal set of chromosomes) that is not known in the wild. The earliest archaeological records of the peanut are from coastal Peru, dated at 3800 BP. Botanists, however, believe that the peanut was first domesticated in southern Bolivia or northern Argentina by the predecessors of modern Arawak-speaking people.[31] The major changes that took place in domesticated peanuts are a shortening of the underground shoots that connect the buried fruit to the mother plant and a thickening of the fruit itself. These changes greatly eased the harvesting of peanuts and made mechanical harvesting possible.

The oldest remains of a cultivated plant in Peru are those of the common bean, found in Guitarrero Cave in the Sierra and dated at approximately 8500 BP. The remains are unmistakably those of fully cultivated plants, not of plants intermediate between wild and cultivated. This is also true of the first finds of other cultivated plants in the area, such as lima beans (the earliest date from about 5300 BP), sweet potatoes (4000 BP), manioc (3000 BP), and potatoes (5000 BP). Is this just a coincidence, or does it have a meaning?

Most evidence of early agricultural activity comes from dry sites. Does this mean that agriculture originated in arid regions? Or that agricultural remains were preserved in dry and not in wet regions? In the Middle East the record is fairly complete, showing most of the intermediate stages between wild plants and their cultivated derivatives. In Peru and to some extent also in Mesoamerica, the intermediate stages are not present. This can be interpreted in two ways. It may be that the intermediate stages have not been preserved or found yet. Another interpretation is that agriculture was not adopted in Peru but imported from outside. The best-known adherent of this theory is the late geographer Carl Sauer.[32] He believed that agriculture in the Americas orig-

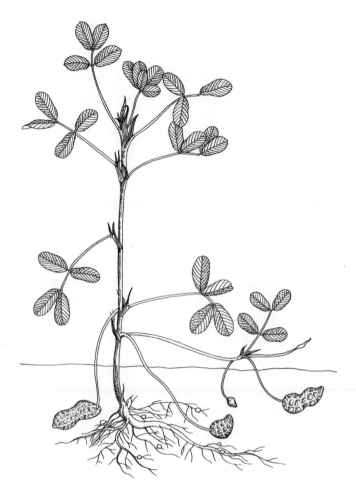

The peanut plant.

inated in the lowland tropics and diffused from there to both Central America and Peru. He suggested this theory not only because people in Peru apparently took up growing fully domesticated plants, but also because several of these species, such as manioc, some wild cottons, pineapples, and peanuts, are native to the American lowland tropics.

Recently scientists discovered 6,000-year-old fossil maize pollen in sediments in the bottom of a lake in Amazonian eastern Ecuador, pointing to an early diffusion of this crop.[33] Lake sediments suggest the existence of agriculture in this area until about 800 years ago. Its abandonment coincides with a climatic change toward increased rainfall known to have occurred about this time. So far no other cultivated species have been identified from that site. It is still too early to tell whether this indicates the existence of extensive agriculture in the lowland tropics.

There are problems with the theory of the lowland tropical origin of agriculture in South America. Not all plants cultivated in Peru or Mexico are from the lowland tropics. Some, such as maize, several species of squashes, and tepari beans, have no South American relative and are clearly of Mesoamerican origin. Others, such as the potato and quinoa, are almost certainly native to the Andes. Moreoever, it is difficult to imagine how humans would have started cultivation in a rainforest environment. The greatest problem with this theory is the lack of evidence of cultivated plants in the lowlands. But even if agriculture did originate there, probably no remains would have survived because of the wet climate. Thus the tantalizing problem persists of why there are no remains of the first stages of cultivation in the Americas.

No one knows whether agriculture originated independently in Africa. Although there are several domesticated plants of African origin, archaeologists have not found any sites showing that people in Africa adopted agriculture on their own. The most likely African sites for the domestication of plants and for early agriculture are Egypt and Ethiopia, although the sub-Saharan belt, west Africa, and even the rainforest areas of central Africa have been considered at one time or another as centers of origin for a number of crops. The most likely explanation is that planting was introduced into Africa from the Middle East.

Grain sorghum and bulrush millet, among the oldest cultivated plants, are the principal cereals of the tropical savannas of Africa and Asia, constituting the staple food of half a billion people. Millet is a name given to several small-seeded grass species. Finger, or African, millet and bulrush millet are important crops in Africa, the former because of its storing quality, the latter because of its resistance to drought.

Grain sorghum, known also as milo or milo maize, is gaining in importance around the world. It is employed for brewing beer and in the United States, Europe, and Japan as animal fodder.[34] Sorghum includes many varieties with different uses. Sweet sorghums have tall juice stems used to produce a sweet syrup. Broomcorn sorghum has dry and rigid stems and produces a stiff, elongated panicle used in the manufacture of brooms and brushes. Sudan grass is an important forage plant used both as a green fodder and for silage. Another species of sorghum, Johnson grass, is good cattle fodder but has otherwise become a noxious agricultural weed.

It is not entirely clear exactly when or where in Africa sorghum originated. It was probably brought under cultivation by the Cushites, a people that 6,000 years ago migrated from the Middle East to Ethiopia, taking with them the knowledge of agriculture.[35] From Ethiopia sorghum was taken across the Sahara to west Africa. It was brought to America with the slave trade.

Cowpeas are another ancient African crop. These highly variable bushy or trailing annual plants are an important source of protein and supplement the staple diet of subsistence farming communities in the savannas of Africa and Asia. Cowpeas are also used as forage plants, as cover plants to reduce erosion, and sometimes as a coffee substitute. Other important crops of African origin are coffee, sesame, watermelons, oil palm, and castor oil.

Archaeologists, anthropologists, botanists, and other scientists have united forces in the last twenty years to study the early development of agriculture. They have proposed a number of detailed theories on how and why agriculture arose, testing them with available archaeological information. Although there is still some disagreement, their research is starting to bear fruit. The principal result, one with which every researcher agrees, is that the adoption of agriculture was a gradual process. It was not a revolution but a gradual evolution. Yet the ultimate impact of agriculture on human history was such that we are justified to speak of a revolution.

At one time anthropologists believed that hunter-gatherers

adopted agriculture because they were half starved. When studies revealed that many contemporary hunter-gatherer societies lead a rather comfortable existence, it became harder to understand why their prehistoric counterparts should have switched so readily to cultivation. What force or forces drove people to adopt agriculture? The exact process must have differed from place to place and over time. One factor seems to be common to all: increasing population. That is, as populations increased and supplies of wild plants became scarce, especially in dry years, people everywhere seem to have supplemented their diet with cultivated plants. However, no one knows whether population increases preceded, coincided with, or followed the adoption of farming.

The reason this process took so long is that farming requires fine-tuned, specific knowledge. Farmers must know how to treat the soil before planting, when to plant, how deep to plant, when to weed, when to fertilize, how to protect the crop from predators and disease, when to harvest, what to harvest, and how to store the crop. Since each crop has its own needs, knowledge of how to cultivate one crop cannot be applied to the cultivation of another. Clearly, it would take many generations of experimentation before humans learned how to grow plants reliably. And given the vagaries of weather, agriculture would never be entirely safe. Its one great benefit was improved productivity. Undoubtedly this was the reason for its adoption worldwide.

Common patterns can be detected at the five prehistoric sites where hunter-gatherers started to grow plants. First, people in all five places had reached a certain level of development in toolmaking, referred to as the Neolithic stage. Second, although these sites were semi-arid, they embraced great ecological diversity in their low valleys, hillsides, high plateaus, and coastal areas. Such diversity may have permitted the coexistence of planting and hunting-gathering. Third, once planting was adopted as a method to supplement hunting and gathering, sedentary villages rapidly developed. In the Middle East sedentariness apparently preceded planting. The populations of settled villages increasingly relied on agriculture for subsistence.

Agriculture, once adopted, spread to other areas. In many of the places, for example regions of Africa, India, and eastern North America, local species were domesticated. These included important crops such as coffee, sorghum, and sugar. An important unanswered question is why agriculture was not adopted in areas such as California and the Argentine pampas, where conditions conducive to farming prevailed.

CHAPTER

Domesticating Plants

CAN ANY PLANT be cultivated? Probably, but certain species are difficult to grow and require special conditions. Seeds of most species have specific germination requirements; seedlings of all species are highly sensitive and die easily from too much water or too little, from too much sun or not enough. Many species require a precise temperature and a certain length of day to bloom and bear fruit. Growing conditions can also affect a plant's susceptibility to insects and diseases. Finally, the right soil conditions are crucial if a crop is to thrive. Some species require heavy organic soil, while others will grow only in loose, sandy soil, and so on.

When farmers plant a seed, tend the emerging seedling, protect it from weeds and pests, and harvest its roots, leaves, fruit, or seeds, they profoundly alter the life of that plant. If they repeat the process year after year they alter forever the evolution of the species to which it belongs. No longer does the plant have to expend energy to defend itself against enemies—farmers will protect it. No longer does it have to compete with other species for space, light, and nutrients—farmers will eliminate competition. Moreover, farmers will allow only those plants that produce many seeds (or leaves, or roots, or fruits, according to the crop) to produce offspring. And not only many seeds, but seeds that are tasty and easy to harvest. The changes that take place in a cultivated plant are profound and rapid.

All today's cultivated crops were critically modified during domestication. Botanists cannot always trace the successive steps involved in modification. Nor can they always determine which are the wild ancestors of cultivated plants, even in those cases where wild relatives are thought to be growing in the same area as their tamed progeny.

Domestication is the process by which a wild plant adapts to the needs of the farmer. The farmer wants species that are nutritious and good tasting, easy to grow and harvest, easy to store and transport. Since the earliest days of agriculture, farmers have chosen species with these characteristics. Within a crop they have propagated individual plants most suited to their needs.

One major result of selection by farmers is that plants no longer drop their seeds when ripe. Wild flowering plants propagate by seed. When seeds ripen they are broadcast into their surroundings in a variety of ways. Some are carried away by the wind, others by animals, still others when the parent plant ejects them at maturity. A plant that does not disperse its seeds at maturity reduces its chances of successful reproduction. From the farmer's point of view, however, a plant that does not shed its seeds before they are harvested provides insurance against the loss of the harvest. If the farmer waits too long before harvesting crops that disperse their seeds naturally, all the seeds will fall off. If the crop is harvested early to keep the seeds from being dispersed naturally, they will be immature. Unripe seeds are less nourishing and usually harder to digest, and owing to a higher water content they spoil easily. They will usually not germinate and therefore cannot be used to propagate the crop. Microscopic analysis of early flint sickle blades shows that the pre-agricultural Natufians of the Middle East harvested unripe cereals.[1] They almost certainly did so because the wild plants they consumed dropped their seeds at maturity.

In most cases, the difference between a plant that drops its seeds and one that does not is genetically simple. The ripening seed (or fruit) is firmly attached to the mother plant. At maturity the cells at the point of attachment, or rachis, dry up and die. When wind or a passing animal rattles the seed it falls off. Several genes control the degree of firmness at the point of attachment. Usually a modification in any of them will determine that the seed remains on the plant. This happens in nature all the time. Wild plants with the modified genes produce few or no offspring. Under cultivation, the opposite is true. Plants with the modification have a greater chance of being harvested and leaving offspring. In this respect, natural selection favors brittleness, while cultivation favors firmness.

The sesame plant illustrates the role of human selection. People have grown sesame since Sumerian times for its seeds, which are an important source of oil. Traditionally farmers harvest the plant before the seeds ripen by cutting it at the base. They hang the plant upside down over a sheet to dry and gather the seeds that fall, thereby selecting for the wild type that sheds its seeds at maturity. More recently, mechanical harvesting of sesame was introduced. Now farmers keep the plant in the field until it ripens, then harvest it by machine. Thus wild plants that shed their seeds are no longer selected. Farmers could quickly select plants that retain their seeds because the difference between a brittle and a firm rachis is genetically simple.[2]

A second widespread result of domestication is the simultaneous or near-simultaneous ripening of a crop. In the wild a species blooms over an extended time. Since the plant cannot predict when the best soil and climate conditions will prevail for its seeds to grow, dropping seeds over an extended period permits some to fall when conditions are best for germination. In a field grown with unmodified wild plants the farmer has to harvest repeatedly during the growing season to capture all the seeds. There is a clear advantage to synchronizing seed maturity:

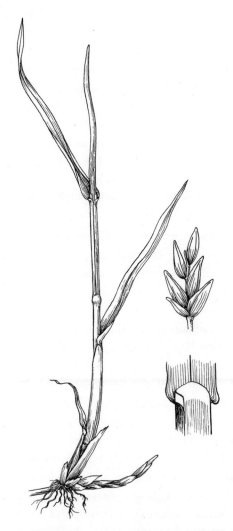

A grass plant. The upright shoot is called a tiller, the lateral shoot is a rhizome. The bottom inset shows the lower part of the leaf sheath that surrounds the shoot and its intersection with the leaf blade. The top inset shows a group of grass flower buds.

It decreases labor and the cost of harvesting. The degree of simultaneous ripening varies with the crop. Some species, such as coffee, tomatoes, and cotton, still bear fruit over a long period and require repeated harvests, but plants in field crops, such as wheat, maize, and soybeans, all ripen at approximately the same time.

A third common change observed in seed crops, especially cereals, is a rise in the size and number of seeds produced. The advantage to people is increased yield. One of the most pronounced changes has taken place in maize, which becomes obvious when a modern corn cob is compared with that of its wild ancestors.

Gigantism, the production of enlarged plant parts, is another common result of domesticity. Potatoes, carrots, cabbages, pineapples, bananas, and apples are just a few of the many examples of crops with enlarged tubers, roots, leaves, and fruits.

A fourth alteration caused by human intervention is a shortened growing season. The less time it takes to produce a crop, the less the chance of a catastrophe such as drought, flooding, or infestation. A shorter growing season reduces the work of tending the crop as well as increases the possibility of planting a second crop in the same field, or of turning the stubble over to grazing livestock. An example of a crop with a shortened growing season is wheat. Wild wheat lives in regions with winter rain and summer drought. It germinates with the rain in the fall, grows through the winter, and ripens in the spring. Without exposure to the cold of the winter it does not bloom. Winter wheats still exhibit this pattern of growth. Spring wheats, which are planted in the spring and ripen in early summer, no longer have the requirement for cold. Some of these varieties are adapted to areas with extremely cold winters, where winter wheats would be killed; others have been bred to grow in the tropics, where there is no winter.

While most wild plants do not produce seeds through self-pollination, many crop species can fertilize their own ovules, which then develop into seeds.[3] This happens in only about a fourth of wild species. Yet even species that are capable of self-fertilization usually crosspollinate in the wild. Crosspollination avoids the adverse genetic consequences of inbreeding. The advantage to the farmer of self-fertilization is that outside agents such as bees or birds are not needed to produce seed. Self-fertilizing plants produce a more reliable seed crop.

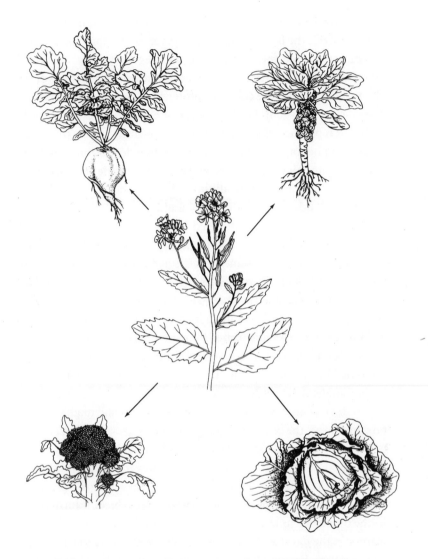

Evolution of plants under domestication as demonstrated by diversity among cabbages and their relatives. In the middle, a representation of Brassica campestris, *wild turnip, and radiating from it, cultivated forms of B. oleracea:* turnip (upper left), brussels sprouts (upper right), broccoli (lower left), cabbage (lower right).

The change to self-fertilization in crop plants is not universal. In wheat and rice the flower produces both ovaries and stamens, and self-pollination and self-fertilization take place regularly. In maize, on the other hand, male and female flowers are borne on different parts of the plant, the tassel and the cob, and self-fertilization, while possible, requires an outside agent. Furthermore, maize seeds that result from self-pollination produce plants that are on the average smaller than their parents, a phenomenon known as inbreeding depression. Some cucumbers, date palms, rye, alfalfa, and many apple varieties still require crosspollination. Farmers usually house beehives near alfalfa fields and in apple orchards to increase crosspollination.

Finally, the tissues of cultivated plants are more digestible than those of their wild relatives. Most wild plants, even edible ones, have substances that when consumed in large quantities are either mildly indigestible or poisonous to humans. For example, broad beans are mildly toxic if consumed in large quantities. Certain enzyme-deficient people who eat raw broad beans or inhale its pollen develop hemolytic anemia, or favism, characterized by abdominal or back pain, fatigue, nausea, fever, and chills.[4] For small children with this deficiency, eating broad beans can be fatal. To reduce the danger of favism broad beans must be thoroughly cooked.

Other crops that retain some toxic compounds and require special preparation are the bitter varieties of manioc, or cassava. They contain poisonous compounds that must be leached out of the roots before being eaten.[5] Unlike the sweet varieties, however, they are resistant to insect attack. Botanists used to believe that they were the primitive manioc, and that the sweet varieties were human bred. It turns out to be the other way around. Itinerant Indians in the Amazon, who plant manioc in the forest and leave it for later harvest, found that the sweet varieties were likely to be destroyed by insects, while the bitter varieties were resistant to attack. They learned how to extract the poison from the latter. At present, most commercially grown manioc is sweet. In Africa, where sweet manioc is an important crop, two insect pests, the cassava mealybug and the cassava green mite, seriously threaten its production.[6]

The poisonous or indigestible substances of wild plants protect them against animals that would otherwise eat them and from disease-causing microorganisms but also make them unattractive as crops. People have therefore bred plants to remove lethal compounds. Since few domesticated plants retain the ability to make toxins, they are more susceptible to pests. Control of plant disease and insect pests is today one of the most difficult and expensive issues in agriculture. Farmers spend millions of dollars each year protecting their crops against disease and pests. For example, in 1986 the United States lost a quarter of its agricultural production to pests and diseases, at a cost of more than $30 trillion.[7] The negative effects of pesticide use have been estimated to cost $955 million a year in the United States. A challenge for modern scientific agriculture is to develop crops that are poisonous or unattractive to insects but at the same time safe for human consumption.

None of today's major crops could survive without human intervention. Wheat is a good example. Humans have been modifying the wheat plant consciously or unconsciously since they started cultivating it—or even earlier, when they were harvesting wild plants. One of the earliest modifications was to select plants that did not shed their seeds at maturity. A second important modification was to select plants whose seeds broke loose from their hulls. An articulation at the base of the seed connects it to the hull. The traditional way of handling spikes of grain after the harvest is to thresh or pound them. Threshing causes seeds to break loose from the hulls. Farmers then winnow the treated grain by tossing it in the air. The lighter hulls (the chaff) blow away, while the heavier seeds fall to the ground. Early on this process selected plants with smaller hulls and brittle articulation.

Although all these changes were advantageous to the farmer, they were detrimental to the survival of crops in the wild. Plants that do not drop their seeds at maturity will not disseminate their seeds to new sites. Seeds lacking distasteful compounds will not protect themselves against predators. And plants with large seeds will usually produce fewer of them, and so leave less offspring.

All these changes occurred spontaneously in some plants, which were then selected by farmers. The process was at least partially conscious.

Cultivators always pick the best plants for their seed. They know that larger seeds not only are more nutritious but also produce stronger plants. Plants that come from large seeds may have genes for large size and may produce some offspring with even larger seeds. Farmers will keep selecting plants with large seeds until they use up all the genetic variation for large size, at which point a crop is unable to produce bigger seeds.

And some selection must have been unconscious. For example, plants that retained their seeds longer had a greater chance of being picked by the farmer, because others had probably shed their seeds by harvest time. Traditional farmers were excellent plant selectors and produced over the centuries many and diverse varieties of plants to suit their needs. Some maize varieties created by traditional Peruvian farmers reaching back all the way to pre-Inca days include popcorn, starchy maize for making *chicha* (beer), and sweet maize for eating fresh. As for potatoes, Bolivian cultivators grow up to thirty different varieties in their fields. Some grow especially well in cold seasons, others in dry or wet ones. There are varieties that produce large tubers, those that produce smaller tubers good for making *chuño* (naturally freeze-dried potatoes), and so on. By growing many potato varieties, Bolivian farmers make certain they have a harvestable crop even in lean years.

For subsistence farmers, the principal objective is to produce a healthy, well-balanced diet. Therefore they select the most nutritious and complementary crops. For example, Mesoamerican cultivators grew a mixture of maize and beans. It turns out that maize is deficient in the amino acids lysine and methionine but rich in tryptophan and cystine, while beans have a complementary amino acid composition. Another requirement for subsistence farmers is stability, the production of a crop year after year. Since crops depend on the weather, farmers tend to grow a mixture of types to ensure some production every year. An anecdote told by Professor Ephrain Hernandez-Xolocotzi illustrates this point.[8] He asked an Indian farmer in Tlaxcala, Mexico, what type of maize he planted. The farmer replied that he planted three kinds: a yellow maize that matured in five months and gave low yields; a cream-colored maize that took six months to mature but

Diversity in cultivated maize. Figure includes some of the largest, smallest, and longest cobs and the corn with the largest grains (Cuzqueño), cultivated in Peru for making chicha, *a corn beer.*

yielded more than the yellow one; and a white maize that was high yield but took seven months to mature. To the question why he did not plant only white maize and thereby maximize his yield, the farmer replied, "Tell me, Mr. Agriculturist, how much and when will it rain next year?" When Professor Hernandez replied that he could not predict the future, the farmer said, "Exactly! Therefore I plant all three types, so if there is a little rain, I always have some yellow maize to eat, if there is more rain, I have plenty of cream maize to eat, and if it is a good year with plenty of rain, I'll have white maize to sell."

In agricultural systems geared for a market, increased yields are the overall objective. Farmers in these systems aim to operate at maximal profit, minimize instability in production from year to year, and reduce long-term degradation of production capacity.[9] The farmer in such circumstances cultivates and selects the highest-yield crops, not necessarily the most nutritious. During the last 2,000 years, market-oriented farming has replaced subsistence farming slowly but inexorably, with yield slowly replacing nutrition and stability as the primary goal of plant breeding.

The birth of genetics at the beginning of the twentieth century made plant breeding more scientific. Understanding how plant and animal characteristics are inherited allowed breeders to accelerate the creation of new varieties and the production of tailormade types. Today we have not only winter and spring wheats but also wheats specially adapted to the conditions of Kansas, of Saskatchewan, of southern England, of Mexico. Breeders learned the factors that determine flowering times in plants and were able, for example, to breed strawberries that bloom throughout the year, thereby changing what used to be a costly and exclusively spring crop into one more affordable and available most of the year. Still, high-yield strawberries with big fruit are not as tasty as their wild, low-yield ancestors.

Breeders have been able to improve the nutritional qualities of plants by breeding for higher contents of certain amino acids and vitamins. However, their main efforts have gone into increasing the yield of crops, their resistance to pests and disease, and their marketing properties, such as endurance and color. The application of new discoveries in molecular biology, often referred to as biotechnology, promises even more and better crops.

An important aspect in plant domestication is isolation from wild relatives.[10] If crop plants can exchange genes freely with noncultivated relatives, and if the latter are more numerous (as was the case in the early stages of domestication), then natural selection might swamp out the results of domestication. According to some scientists, humans would not have had any incentive to cultivate in the same place where plants grew spontaneously.[11] Presumably crops were planted where they did not grow wild, and as a result they became isolated from their relatives. Because pollen generally travels only short distances, growing plants in a plain near a camp rather than on their native site on a hillside could have resulted in isolation. Farmers also isolated plants from wild relatives by selecting self-pollinating varieties.

Up to the middle or late nineteenth century all crops were the product of generations of selection and inbreeding mixed with hybrids, the product of random crossings in earlier generations. In many places, especially where farmers practice traditional agriculture, these land races still prevail. But where modern agriculture is practiced, pure-line, high-yield varieties that are much more uniform genetically and much more demanding in their growing requirements displace land races. The result is a narrowing of the genetic base of most if not all major crops, a process known as genetic erosion.

There is an interesting distinction between the type of agriculture originally practiced in Europe and northern Asia and that practiced in the Old World tropics and in the Americas.[12] In the first kind of agriculture, farmers cultivated species that have edible seeds such as wheat or rice in open fields with the help of plows. In the tropics farmers used hoes and cultivated species with edible roots or tubers, such as yams, manioc, and taro. In Mesoamerica and Peru local farmers cultivated both seed crops and root crops. Native American farmers grew their crops with the help of hoe and digging stick; the plow was unknown until the arrival of the Spaniards. And they grew plants with large seeds, such as maize, beans, and squashes, while the farmers of the Old World grew smaller-seeded plants, such as barley, wheat, millet, and lentils.

These two types of agriculture, called seed crop cultivation and vegeculture, respectively, have different ecological impacts.[13] Most seed crops are (or were before the invention of mechanical seed drills)

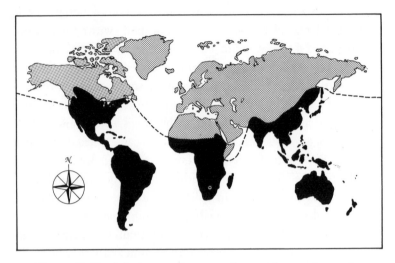

*The so-called hoe belt, the area where vegeculture predominated around
6000 BP. North of it, seed culture using the plow was the norm
(according to Braudel).*

planted by broadcasting the seeds into large fields freed of vegetation
with the aid of a plow. Open fields are prone to wind and water erosion.
Moreover, farmers plant only one seed crop in each field. This tends to
exhaust the soil, requiring the farmer to move to another field or rotate
crops and fertilize soil. And a field with hundreds or thousands of plants
of the same species is very susceptible to insect attack and disease.

In vegeculture, the crop is propagated by cuttings or some other veg-
etative means. (In American vegeculture farmers also cultivated some
seed crops, such as maize and beans.) Farmers clear their fields of only
the most obnoxious weeds and plant their cuttings in holes made by a
hoe or digging stick. Clearing the entire field would add unnecessarily
to the labor of the farmer. Since farmers cultivate each plant separately, they can grow a number of crops in the same field. This reduces
soil depletion and protects the field from erosion, since there always is
a cover of vegetation. Furthermore, the coexistence of different crops
reduces the likelihood that one disease or one specific insect will destroy all the harvest.

Vegeculture requires more individual labor. Farmers cannot mechanize it as readily as seed-crop agriculture. It is therefore quickly disappearing. Today farmers plant even root crops such as potatoes or manioc in monocultures.

How agriculture changed the human diet is a complex subject. It is difficult to find out what people ate in the past, given the perishable nature of food. Existing evidence indicates that the diet of pre-agricultural people was diverse, and that over the course of human history diets became more uniform and less distinct.[14]

Humans are omnivorous, that is, they eat both plant and animal food. Given a choice, and unless they have some ethical or religious scruples, people prefer animal flesh over plant material (although most people are primarily seed eaters). This is probably because animal flesh has a higher content of protein than plant material. Also, animal proteins are more similar to human proteins than are plant proteins, and therefore humans assimilate them more easily. Of course, people in the early days had no idea what proteins were, and even today we prefer meat not because it has more protein but because we think it tastes better. Studies show that all animals, from insects to ruminants, have definite priorities when given a choice of what to eat. Although it appears to a casual observer that a grazing cow or horse is eating any and all the grass in a field, this is not so. Grazing animals carefully choose among the species growing in a pasture. Their highest priority seems to be the quantity of protein in food, although apparently they also balance factors such as digestibility and the amount of silicon. Other animal species are also discriminating eaters, as any pet owner knows. This is curious, since animals have no sense of how much protein is contained in the food they eat. What we call taste is a mechanism that has evolved over time by means of which an animal unconsciously chooses the most nutritious (or close to the best) available diet.

The first dietary priority is to obtain calories. A calorie is a unit of energy contained in a food. Bodies burn most of the food they eat to produce energy. Humans get most of their calories from starchy or sugary foods obtained mainly from plant products such as cereals and potatoes. Fats and proteins are also a source of calories. Fats have twice the calories per gram that starches have. The number of calories a person needs varies according to sex, age, size, level of activity, and climate. A man who weighs 132 pounds (60 kilograms), engages in moderate activity, and lives in a temperate environment requires about 2,600 calories per day. A man weighing 220 pounds (100 kilograms) under similar circumstances requires 3,700 calories.

We also use part of what we eat as building blocks for our bodies. Our own cells synthesize most of the compounds we need from simple molecules and mineral atoms contained in food. Some compounds, however, have to be consumed. Among these are some amino acids (the building blocks of protein) and vitamins. We need smaller amounts of the latter than of the former.

Proteins are the basic fabric of our muscles and the fleshy part of our bodies. Meat, especially that of mammals, contains all the needed amino acids and usually in the needed proportion. Plant proteins differ from animal proteins, and depending on the species might lack one or more amino acids essential for humans. A gram of maize contains about a third of the protein in one gram of milk, but less than a sixth of the amino acid methionine. Thus, although three grams of maize provide the same amount of protein as one gram of milk, it takes six grams of maize to provide the same amount of methionine. A diet based on just one plant product, such as maize, is inadequate and leads to serious malnutrition.

Only plants synthesize vitamins (there are a few exceptions, such as the synthesis of vitamin D in human skin).[15] Carnivorous animals either get enough vitamins from their prey (which obtain them from plants) or supplement their carnivorous diet with just enough plant material to fulfill vitamin needs. People also obtain vitamins from the food they eat. Yet when their diet is not diverse enough, especially if it is deficient in fresh vegetables, they show symptoms of vitamin deficiency.

Diseases resulting from vitamin deficiency are not as common as they used to be. Beriberi, which results from a shortage of vitamin B_1 and is characterized by muscular spasm and heart damage, used to be common in China, particularly after the introduction of polished (dehusked) rice. Scurvy, resulting from a shortage of vitamin C, produces hair, nail, and tooth loss, muscular atrophy, and death. It used to appear so regularly in winter months in England that people called it the London disease. A deficiency of vitamin D, known as rickets, occurs most often in northern climates, where there is less sun. These nutritional problems are in part the consequence of depending on the harvest of a limited number of species.

Seeds are among the organs of crop plants that farmers have modified most. The reason is that they constitute the most important part of the diet of agricultural people. Although humans are omnivorous and eat many kinds of plant and animal products, most are primarily seed eaters. The importance of seeds in the human diet is not obvious because they are consumed in the form of bread, pasta, oil, and breakfast cereals. Even so, we eat rice, maize, beans, peas, lentils, peanuts, and different kinds of nuts with little elaboration. Most large mammals are either flesh eaters (wolves, cats) or leaf eaters (horses, cows, elephants). Why are humans primarily seed eaters rather than leaf eaters?

Seeds contain all the nourishment (starch, proteins, vitamins) needed by the seedling during the first days of its life. This makes seeds very attractive as a source of food for animals and people. The type and quantity of reserves and the chemical composition of seeds can vary widely. They contain mostly starch, but some, like peanuts or sunflowers, hold significant quantities of oils. Others, such as soybeans or sesame seeds, contain proteins. Some seeds are also rich in vitamins. Grass seeds contain mainly starch and some protein. Of the twenty most widespread cultivated crops, eleven are grown for their seeds, three for their fruits. Of these fourteen, six belong to the grass family and three to the legume family.

Seeds have another characteristic that accounts for their importance in agriculture. They are dry and will keep for months and even years if protected from insects and rodents. Because seeds store so well, they made human survival possible during droughts and cold winters. Because they store well and are also easily transportable, they can be used for trade. Seeds range in size from those of orchids, not much larger than a speck of dust, to the double coconut, a seed from the Seychelle islands that can weigh two kilograms (four pounds). Most seeds are small, weighing only a few grams, and are disseminated by the plant as soon as they are ripe, making it hard work to collect them.

Some underground organs, such as potato tubers or manioc roots, are rich in food reserves, and for similar biological reasons. The food reserves of underground organs provide nourishment to the plant's offspring in the early developmental stages. Roots and tubers have the advantage over seeds of being larger and in one place. Yet roots, tubers, and fruits are bulkier and much more perishable than seeds, and people must consume them shortly after harvest. The same applies to leaves such as lettuce. Underground organs are also harder than seeds to harvest.

When offered a choice, large herbivorous mammals such as horses and cows usually prefer seeds over grass. Race horses consume a diet rich in seeds, primarily oats. But most large mammals normally eat grass. Why is that, if seeds are so nutritious? It is because seeds are seasonal, and large animal species without tools and adequate mobility cannot find enough seeds to sustain them throughout the year. Even hunter-gatherer humans could not live exclusively from seeds.

Nobody knows exactly how many species of plants inhabit the earth. There are at least 300,000 species of terrestrial plants, most of them growing in the tropics. Botanists classify these species into some 400 families. Families range from very small ones with one or a few species, such as the inconspicuous Halophytaceae found only in Patagonia, to the common and beautiful orchid family (Orchidaceae), one of the largest families of flowering plants, with over 20,000 species. But of the thousands of plants that people have grown over the centuries, they domesticated fewer than 200 species for food or fiber. Twenty of these species form the bulk of agricultural production worldwide.

Of the twenty most widespread cultivated species, eight belong to the grass family. They include the world's three most important cultivated species, the cereals wheat, rice, and maize. The grass family is the fourth largest and the most widespread family of plants. There are between 9,000 and 10,000 species of grasses, about as many kinds as there are of birds. They grow in all types of habitats, from the Arctic to

Antarctica, where one of the two species of flowering plants is a grass. Grasses grow in mountains and in forests, but they prefer open space with a lot of light and few trees, such as steppes and savannas. Grasses come in a great diversity of sizes, from giant bamboos to small cushion plants only a few centimeters tall. In spite of such variations, the basic construction of the grass plant is consistent. One does not have to be a botanist to recognize a grass.

The stem or shoot of a grass consists of a series of hollow tube segments separated by solid disks from which the leaves emerge. We call the disks nodes, the hollow tubes separating them internodes. The leaves of grasses look very much alike, at least superficially. They consist of a basal area called the sheath, a blade, and a thin membrane, the ligule, separating the first two. The sheath envelops the internode immediately above the node from which it emerges. The blades emerge at an angle from the sheath and are flat and elongate.

The flowers of grasses are unlike those of any other plant type. Grass flowers have no petals or sepals. Instead they have two modified, boat-like leaves or bracts that form a kind of box, hinged at the base. These open at maturity, exposing the pollen-bearing stamens and the feathery stigma, and later surround the developing seed. When the seed is ripe, the rachis, or stem that holds the seeds and bracts together, breaks below the bracts, and the seeds, surrounded by their bracts, drop to the ground.

Grass flowers form at the end of the shoots, in groups or bunches that botanists call inflorescences. These can be very loose, as in rice and oats, in which case botanists call them panicles, or very tight, as in wheat and barley, in which case they are called spikes. After a grass shoot blooms it—but not necessarily the entire plant—invariably dies. Many species of grasses live less than a year, such as our cereals, but many more live for several years, such as the grass in a lawn. Some grasses, like bamboos, bloom infrequently, once in decades, after which the entire plant often dies. Most grasses, however, bloom every year; after that the flowering shoots die and are replaced by new shoots that grow from the base of the plant. This gives most grasses their globose, bunchlike shape.

The word *cereal* comes from Ceres, the Roman goddess of agriculture. Cereals have some common characteristics. All are grasses cultivated for their seeds. Wheat, barley, and rye are closely related and their spikes look alike. Maize and sorghum are related, but not as closely. Cereals are all annual species, which means that farmers must plant and harvest them each year. Cereal grains, which we have been calling seeds, are botanically speaking fruits. They are hard and dry and have to be treated before they are edible. Wheat, maize, and rye grains are normally ground to produce flour. Oats, millets, and sorghum grains are usually soaked and made into some kind of porridge. Cereals can also be popped by heating. Ancient Egyptians popped barley and wheat, and today people eat popcorn. Rice seeds are generally dehusked and boiled. All these methods have been used to make the hard seeds of cereals more palatable.

Pulses, or legumes, members of the family Leguminosae, are second only to cereals as sources of human food. Seeds of legumes are especially rich in protein because most legume roots have a symbiotic relationship with bacteria, which take nitrogen, an important component of protein, from the air. This symbiotic association allows legumes to produce more protein than most plant species. Legumes, however, have poor bulk yields compared with cereals and even poorer compared with root crops.

The family Leguminosae includes not only herbs but also many shrubs and trees. It is the third largest family of flowering plants, with more species than the grass family (about 12,000), and it is particularly important in the tropics. As we have seen, many legumes are a significant source of human food. Others, such as alfalfa, clover, and vetch, are important forage plants. Many tropical trees in this family are valuable for timber. Finally, legumes are important sources of gums (gum arabic), dyes, and fibers.

The ancestors of many of the cultivated pulses are either slender annual plants or small shrubs. Like cereals, domesticated pulses differ from their wild ancestors in characteristic ways. The most widespread change parallels one found in cereals: At maturity seeds are not dispersed. Wild legume seeds are contained in fruits called pods. At maturity the pod breaks open, often explosively, and scatters the seeds. This is not the case in most cultivated pulses. Another characteristic change is a shorter and more robust stem. For example, field-grown pea plants are short, stocky, and shrublike, while garden varieties have to be propped up. The wild ancestors are slender climbers with long shoots.

Other families that have given us important crops are the Cucurbitaceae (squashes, pumpkins, watermelons, melons, cucumbers), Solanaceae (potatoes, tomatoes, chili and green peppers, eggplants, tobacco), Rosaceae (apples, pears, plums, cherries, strawberries), Liliaceae (onions, garlic, shallots, leeks, asparagus), Rubiaceae (coffee, quinine), Rutaceae (citrus fruits), and Malvaceae (cotton, okra).

The adoption of agriculture was a revolution in many ways. Humans became dependent on a small number of plants and animals for the bulk of their sustenance, gradually abandoning the diverse diet of their hunter-gathererer ancestors. What we don't usually realize is that domesticated plants and animals became equally dependent on humans for their survival. This is a curious and fascinating relationship. It is not, however, a closed circle of cause and effect. Both humans and the species they domesticated also rely on the environment—sun, air, soil, and water—for their existence. And the environment, to some extent, depends on them for its health. There are no isolated phenomena in nature: The closer we look, the farther the web of interdependence seems to stretch.

The Rise of Civilization

THE DEVELOPMENT OF crafts, art, religion, science, architecture, and all the richness and complexity of human activity that we call civilization was not possible before the rise of cities.[1] Cities arose first in those regions of the world where people adopted agriculture: in China, India, the Middle East, Africa, and the Americas. Their growth was slow: Several thousand years elapsed between the time people began practicing agriculture and the appearance of cities with monumental architecture, organized religion, and rulers and courts.

Cities evolved from agricultural villages. However, a city was not just a large village. It was a seat of government, crafts, and organized religion. It was home to artisans, soldiers, bureaucrats, priests, and rulers. Farmers had to produce a surplus of food to feed all these nonfarming inhabitants. The change from farming for subsistence to producing a surplus for sale, barter, or tribute represents a change in the social organization of farming communities that is almost as radical as the adoption of farming itself.

One important step in the direction of civilization was a sedentary life-style. Unlike nomadic hunter-gatherers, who acquired fresh provisions on a daily basis, farmers did not produce a steady supply of food. A large amount of a limited number of cultivated species ripened at certain times of the year. With its rhythm of planting and harvest, agriculture made the storage of crops mandatory. Crops had to be protected against marauding neighbors and stores managed so that everybody would be fed until the next harvest was in. A fortified village was the ideal place to store reserves.

The first cities emerged quite some time after sedentary agriculture began. The reasons are clear. Cities needed a steady and dependable supply of food to function. Without a cheap and rapid means of transport, food had to come from the area surrounding the city, which had to produce a reliable agricultural surplus. The city was not separate from the country in today's sense; its life was intimately bound up with that of the countryside. Many city dwellers regularly tended their fields outside the immediate boundaries of the city. Nor was agriculture originally an activity performed by one group of people, farmers, to supply another group, city dwellers. Rather, it was the main activity of all people. However, those who controlled the storing and distribution of the harvest soon learned that these responsibilities gave them a great deal of power. As time went by and societies became more hierarchical, those who toiled in the fields to produce reserves were excluded from the processes of allocating them.

Ironically, those at the lower end of the socioeconomic ladder who toiled in the fields contributed only marginally to the evolution of agriculture. People who wielded power in the city usually owned the land, and they were the ones who advanced farming techniques, under the pressure of economic forces generated by the city. Historians are not totally inaccurate when they see history as emanating from cities.

What we must remember is that the ancient city did not end at the city limits but encompassed all the surrounding agricultural land that supplied it.

We have touched on the intriguing historical fact that a cereal is associated with the evolution of all major classical civilizations. The rise of Western civilization, for example, can be traced to the domestication of barley and wheat in the Middle East by ancestors of the Sumerians and Chaldeans. The rise of Mesoamerican civilizations and that of the Incas in South America is linked to the domestication of maize in Mexico. Why are grasses specially qualified to be the primary sustenance of classical civilizations?

Grass seeds are nutritious and they keep well—essential conditions for any stored food that must feed people for at least a year. Root crops and fresh fruits are more perishable and not as nutritious. Grass seeds have a good balance of starch and protein and a decent amino-acid composition. As we have seen, most grass seeds when eaten with a legume make a fairly balanced diet. Grass seeds are also easy to transport. Armies, Napoleon once said, move on their stomachs. For hunter-gatherers, warfare could consist only of small raids on neighbors across country that provided adequate sustenance. With a store of grass seeds, however, their soldier descendants could march for days or even weeks across barren land. Moreover, because grass seeds store so well they are conducive to trade. Regular trade requires a steady supply of seed; simple gathering cannot provide reliable supplies, but cereal agriculture can. For all these reasons, cereals are ideally suited for the role they play as the principal agricultural surplus.

As early farmers became sedentary in order to tend their cereal crops and promote trade, new problems arose. A store of food, particularly at harvest time, must have been a temptation for roaming nonagricultural bands. Successful defense required organization and some kind of leadership. Defense was easier if everybody was living together in some manner of fortified hamlet. Then further organization and more centralized leadership were needed to keep people living and working together in a settlement.

Thus cereal agriculture demanded a modicum of the knowledge and organization we associate with civilization. A leaf crop such as spinach or lettuce would not have made the same demands of a farming community. An intriguing question is whether cereal agriculture begot civilization, or whether incipient civilization invented cereal agriculture.

Cultivation of food plants, especially cereals, was established over most of the humid mountain valleys of the Middle East by 9000 BP.[2] Hulled two-rowed barley and einkorn and emmer wheat, both wild and cultivated, were the most important elements in the diet, followed by cultivated lentils, peas and vetch, and wild pistachio nuts and acorns.

Yet more than 3,000 years would elapse before the first well-defined civilization would arise in this region, that of the Sumerians, in the lower regions of the Euphrates River in what today is Iraq. Since the Sumerians left behind a written record, much more is known about them than about other groups of people then forming civilizations.

The communities in the Levant, Anatolia, and Mesopotamia that first practiced agriculture did so by adding artificially raised crops to a hunting and gathering food economy. This first swidden agriculture was simple. It consisted of scratching the surface with stone axes or primitive hoes, sowing, and reaping. Farmers cultivated their fields for not more than two or three years, then cleared new ones. Eventually when there was no virgin land left, they cultivated again the fields that had lain fallow the longest. Yields must have been low at the beginning. However, as time went by people improved the productivity of their fields by using fire to eliminate unwanted vegetation, animal manure to improve soil fertility, and improved seeds, among other measures. The principal advantage of agriculture at this stage was that fields bore plants more useful than native vegetation.

The sort of shifting cultivation they practiced was sustainable as long as there was enough land available. Villages lay far apart. In the mountains of the Fertile Crescent, where places conducive to cultivation were small and scattered, probably a single village occupied each valley. The ecological heterogeneity of these mountain valleys also offered plenty of wild species of animals and plants for hunting and gathering. For over 3,000 years, from the time of the domestication of cereals and sheep and goats until the occupation of the lower Euphrates Valley, the inhabitants of mountains of the Fertile Crescent occupied their settlements for a few hundred years and then abandoned them. A possible explanation for the departure of the inhabitants from these settlements is soil exhaustion. Charred buildings, crushed skulls, and other evidences of strife at archaeological sites also suggest conflict for land between settled peasants and wandering nomads. The increasing number of settlements as time went by reveals that the pastoralists and agriculturists were gaining the upper hand.

As the population increased, farmers cleared more and more forests for arable land. Their slash-and-burn agriculture was very wasteful of land. Soil erosion worsened as loose sediments were washed away by the winter and spring rains to be deposited in the lower floodplains of the rivers, the Persian Gulf, and the Mediterranean Sea. This erosion has contributed to the Euphrates and Tigris deltas extending 130 miles into the Persian Gulf over the last 5,000 years.

As farming communities in the mountains expanded and used up all suitable soil, populations migrated in search of new farmland. About 6,000 years ago some of these people settled in the lower Euphrates Valley at a place known as Eridu. The valley was a flat, arid, rather sandy plain. The Euphrates had a rich supply of fish and waterfowl that probably made up the bulk of the original settlers' diet. At the time of this early occupation the river overran its banks during the spring flood, creating an area of marsh surrounded by desert. The soil was soft and easy to cultivate. If farmers planted seeds soon after the spring flood these would germinate and grow but usually not mature because of lack of rainfall. The inhabitants solved the problem when they created impoundments or dikes to store water for their crops during the growing season. At Eridu, fields were originally irrigated by diverting water upriver, then letting it flow down the gently sloping terrain toward the fields.

After irrigation was adopted, the villages in the region expanded. The rivers provided plenty of fish and waterfowl, the rich riverine sediments deposited annually did away with problems of land exhaustion, and the grasslands and bayous provided fodder for herds of domesti-

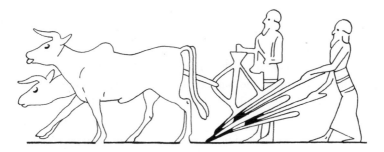

Babylonian scratch plow pulled by oxen with seed drill. From a cassite cylinder-seal, second millennium B.C.

cated animals. Barley and emmer wheat were the first crops planted in this area.

The stone-free terrain of the lower Euphrates Valley made plowing with donkeys and oxen possible. In the stony hills and mountains hoeing was the better method, but it demanded a great deal of labor. Plowing increased labor productivity. It broke up compact soil, allowing water to penetrate deeper. Plowing also eliminated weeds, at least temporarily, and increased the chances of seed germination. It did bring with it two disadvantages. Breaking up soil disrupted the microflora and microfauna that play an important role in the cycling of soil nutrients. And loose soil was more prone to erosion by wind and water. The latter became a serious problem in Sumeria and to some extent has afflicted every plowed field since that time.

The dependence on water and the fertilizing effect of sediments deposited in the spring flood meant that the population had to concentrate along the river. No longer could they disperse over a large area. The denser population no doubt intensified conflicts between the diverging interests of fishermen, herdsmen, and farmers. The need to adjudicate disputes led to the emergence of a class of managers with authority to resolve conflicts. Moreover, water supplies could be cut by enemies or natural catastrophes. Control of a canal by an unfriendly tribe several miles upstream could doom the food supply of an entire people. Unusually high floods could clog or destroy canals or wash away

fields. Defense of the canal system against attack, as well as maintenance of the works in operating condition, called for some form of centralized authority.

The local peculiarities of desert riverbanks help to explain the shape
of social evolution in the Sumerian and other Mesopotamian communities.[3] For example, the rivers served as natural avenues of movement in trade. The barren plains were short of stone, wood, and metals
needed to build houses and make tools. This stimulated plains people
to exchange their farming surpluses with the inhabitants of the hills for
construction materials. One result was that Sumerians became good
shipbuilders. Their ships made out of reeds may have sailed as far as the
Indus River Valley in Pakistan.

Sumerian civilization was city-based. We have mentioned Eridu.
Lagash, Uruch, Ur, and Babylon were also important communities.
At the beginning cities were little more than overgrown villages, most
of their inhabitants being farmers. However, the peculiarities of irrigation farming required a much higher degree of organization than in
the mountain villages of early Neolithic agriculture. Rather than each
farmer tending to his fields, Sumerians collectively cultivated large
tracts of land "owned" by individual gods and administered by priests
on the gods' behalf. One or more such temple communities formed a
city. The priests served as managers, planners, and coordinators of the
large-scale effort to cultivate fields and maintain irrigation systems.
They kept boundary markers, allocated land, and supervised the storage of surpluses in temple granaries. Without such careful planning
and coordination, Sumerian and later civilizations could not have
come into existence.

It was in many respects a simple life for the great majority of people.
The Sumerians made use of some copper, bronze, and iron implements. But metal was a rarity. Most of their tools were of stone; some,
such as sickles to cut cereal, were made of clay. Needles and awls were
made of ivory or bone. Houses were built of reeds and usually plastered
with a mixture of clay and straw. Doors, made of wood, had stone
socket hinges. Floors were beaten earth; roofs were reeds, often covered with mud. Cows, sheep, goats, and pigs roamed freely around the
house.

By 4000 BP the Sumerians had extensive irrigation works branching out from the Euphrates. The Euphrates and the Tigris brought their sediments in large part from the mountains of Armenia. The rate of erosion in these mountains increased steadily during the first 3,000 years of agriculture because upstream farmers destroyed the native forests. For the downstream farmers this meant an endless struggle to keep irrigation canals free of silt so that life-giving water could flow freely to the fields.

As the population grew, the cultivated area and number of canals and ditches increased along with the complexity of the task of keeping the irrigation system working and free from enemies. The development of this system was a remarkable achievement of mind and muscle. A vast maze of waterways laced the flat valleys. A network of dams and channels controlled the flow of irrigation water. The main irrigation canals were lined with burned brick and the joints were sealed with asphalt. At its peak, the system may have irrigated 10,000 square miles of cropland.

The agricultural community of Mesopotamia planted what has been called the Middle East complex of crops.[4] The staples were barley and einkorn and emmer wheat. It is likely that by the time Sumerian civilization arose, common wheat and rye had been added to the list of cultivated cereals. Less likely was the cultivation of oats at this time. Several pulses would have figured among the plants in Babylonian fields: lentils, peas, chickpeas, and broad beans. Flax was no doubt grown both as a fiber plant and for its edible seeds. Farmers cultivated grapes and made wine and brewed beer. Another plant under cultivation by that time was the date palm. People supplemented their diet with fish from the river and meat from domesticated animals—goats, sheep, pigs, and cattle. No doubt people also gathered and consumed edible wild plants and hunted game.

Assyrians eventually overran Babylon and the other Mesopotamian cities. They were followed by a long list of conquerors and empires. For some 4,000 years the inhabitants had struggled with the river to keep canals from clogging and water flowing to the fields. Periods of neglect were followed by periods of reconstruction. But slowly, inexorably, the endless shoveling of mud from canals created huge mounds of silt and muck. People dug new canals that eventually suffered a similar fate. Centuries of flood irrigation led to the accumulation of salt in the fields and forced their abandonment for newer ones. Also, the Euphrates had built up its channel so high with silt that it was constantly changing course. Each course change meant digging new canals. To this day the river still brings its water and its silt, but a much smaller surface area is available for cultivation. The ruins of the ancient cities sit in the desert surrounded by sand and desolation.

The other great civilization of antiquity was Egypt, also closely connected to floodplain agriculture. Historians divide the history of ancient Egypt into two sharp periods: before and after the establishment of the first centralized kingdom in 5100 BP. The early history is incomplete, but the story of the Egyptian kingdoms is fairly detailed thanks to the large number of written records left by the Egyptians, their monuments, and the thousands of elaborate tombs they built.

The Nile has its origin in the great lakes of central Africa and in the mountains of Ethiopia. In its trek north, it flows first through savannas and gallery forests. When it enters northern Sudan and Egypt it meets one of the most extreme deserts in the world. The lower valley of the Nile from the present Sudan-Egypt border north to the Mediterranean Sea is a giant oasis surrounded on all sides by inhospitable desert. When the river reaches the Mediterranean it branches into a large and fertile delta region. The desert protected the early agriculturalists who settled here from attacks by nomads, a situation in sharp contrast to that which prevailed in Mesopotamia. Moreover, the northern flow of the river combined with prevailing southerly winds allowed free communication by boat along the Nile.

Little is known about the early settlement of northern Africa and the Nile Valley and the beginning of agriculture in this region. Better and more direct evidence of crops is needed to fill in the gaps. Agriculture may have been independently discovered by people in central and western Africa, who domesticated some of the plants in common use today, such as sorghum. However, no one knows whether ancient Egyptians learned farming from their brethren farther south, from people living in the Fertile Crescent to the east, or on their own.

At the end of the Pleistocene, northern Africa endured a colder climate than today. This allowed the flora of the Mediterranean region, including the Middle East, to move into northern Africa. One species might have been barley. The milder weather also allowed people to settle this area and, according to some researchers, domesticate wild cattle, goats, and sheep.[5] People established settlements along the river, where a mixed economy of hunting, gathering, and fishing prevailed. Sometime between 7000 and 6000 BP people started growing crops. This coincided with a drying period when there would have been a reduction in the supply of game and wild plants and with the appearance of polished stone (Neolithic) tools. They had tools for grinding seeds, building boats, making pottery, and weaving. Agriculture led to a growth of villages along the river. Just as in Mesopotamia, the higher concentration of people with diverging interests (fishermen, farmers, herders) must have created conflicts and the need for institutions to resolve them. And as the society grew the need must have been increasingly felt for artisans who specialized in the making of agricultural tools, ceramics, boats, textiles, and so on.

The emergence of class stratification in Egypt did not follow the Mesopotamian pattern. Whereas in Mesopotamia social and economic differentiation led relatively quickly to the development of independent craftsmen, trade, private ownership of land, and the emergence of city-states, in Egypt the process not only took much more time but occurred in a quite peculiar, characteristic way. One reason might be that agriculture in the Nile Valley was unique owing to the yearly flooding of the river. Almost 3,000 years after the rise of Egyptian civilization, in 450 B.C., Herodotus commented as follows on the agriculture of ancient Egypt: "They gather in the fruits of the earth with less labor than any other people, . . . for they have not the toil of breaking up the furrow with the plough, nor of hoeing, nor of any other work which all other men must labor at to obtain a crop of corn; but when the river has come of its own accord and irrigated their fields, and having irrigated them has subsided, then each man sows his own land and turns his swine into it; and when the seed has been trodden into it by the swine he waits for harvest time; then . . . he gathers it in."[6] This picture is not totally accurate. For example, ancient Egyptians considered pigs unclean animals and therefore probably would not have used them in planting. There are, however, depictions from the pyramids at Giza, 3750–3600 BP, of farmers using goats to bury the seeds being sown on still-damp fields. Fields had to be worked, but plowing the soft mud and silt was indeed less arduous than plowing or hoeing the hard fields of Greece. Overall, the amount of work needed to farm in the Nile Valley appears to have been considerably less than in the lower Euphrates. The Nile's yearly flooding irrigated and fertilized the soil, preventing the need for an elaborate irrigation system. The inhabitants of predynastic village communities in Egypt therefore had ample leisure time. They used some of that time to develop specialized crafts such as weaving, carving, pottery, stoneworking, and, in upper Egypt, metalworking. There was also ample land so that territorial conflicts were fewer than in Mesopotamia. These ecological factors explain, at least in part, why ancient Egypt did not develop an indigenous, centralized, city-based social structure. The agricultural village continued to be the economic center.

Most traces of life during the earliest periods of lower Egypt are irretrievably buried beneath the alluvial silt deposited by the Nile. The earliest records of fishing and farming villages come from the Lake Fayum depression, dated at 7000 BP. Subterranean granaries, straight-handled and flint-toothed wooden sickles, and threshing flays attest to the use and storage of cereals. The few recovered seeds indicate cultivation rather than gathering. In addition, remains of woven linen cloth give evidence of the growing of flax. From 6000 BP there are traces of agricultural villages at Merimdeh ben Salama on the fringes of the delta. Archaeologists have uncovered additional sites from this time in upper Egypt. The inhabitants produced elaborate flint tools, made fine pottery, and had developed metallurgy. They practiced cereal agriculture, and there are indications that the grain was not only boiled but also baked into bread.

Although the Nile was protected by the desert, invasion was possible from people living in the savannas to the south and from across the Sinai Desert to the east. One invader from the south unified all the scattered settlements along the river some 5,000 years ago, when most of the Nile Valley was largely unreclaimed marsh. Egyptian civilization arose quickly after the establishment of the first royal dynasty at about 5100 BP.

Egyptian art and civilization developed primarily around the pharaoh's court rather than a large city. The pharaohs built the monumental pyramids and palaces with which we are all familiar, and thousands of artisans worked for them to produce some of the most spectacular art ever made by humans. Most of this sprang from their cult of the dead, not agricultural cults.

No river on earth has influenced a civilization and a people more than the Nile. Before the construction of the Aswan High Dam, the Nile faithfully overran its banks every year, depositing a layer of rich and fertile silt on its banks. The Nile drains much of central east Africa and the Ethiopian Mountains. This is an area of seasonal rains that start in April and end in October. In upper Egypt the Nile reached its lowest point in June, by mid-July started to rise, and by mid-August reached its halfway mark. In late September it started to flood the fields, reaching its highest level by the beginning of October and staying there for about two weeks. After that it dropped, at the beginning slowly and then faster until May or early June, when it reached its lowest level. The river controlled the three seasons of ancient Egypt, each about four months long. *Sa* or *se* was the time of the flood, *per* the time of planting, and *semu* the season of the harvest. When the Nile flood dropped below normal, famine invariably followed. When the flood was higher than normal, the river was likely to deposit less silt, reducing soil fertility and crop yields. Not surprisingly, ancient Egyptians attached godlike qualities to the Nile and made offerings to it.

The river inundated only the area close to its banks. Fields farther away had to be watered by an elaborate system of canals. Likewise, farmers had to irrigate their crops since the moisture left by the flood did not last through the entire growing season. Originally farmers used large wooden hoes to cultivate. The stoneless fields led to the early development of a wooden plow drawn by oxen or horses. Farmers used sickles to take in the harvest.

Cereals were the mainstay of ancient Egyptian agriculture. Archae-

*Ancient Egyptians plowing with the help of oxen. Harvesting grain with
the aid of a sickle. From a tomb painting.*

ologists have found seeds of common hexaploid wheat in ancient Egyptian tombs dating back to 5000 BP. Egyptians used wheat primarily for breadmaking (there are numerous and detailed depictions of bakeries in murals). It was also the principal product of commerce in classical times, first with ancient Greece and then with Rome. Egyptians also planted barley widely, using it as food and for wine. They made beer in industrial quantities. Crops of a more horticultural nature were widespread—onions and garlic, bottle gourds, asparagus, legumes including broad beans and lentils, radishes, carrots, okra, cabbage, chicory, watermelons, and dates.

In addition to food crops, ancient agriculturalists also grew fiber plants. From earliest times people used plant and animal fibers for clothing, shelter, and tools such as baskets and nets. People probably collected and used plant fibers long before the invention of agriculture. They used them for making ropes, nets, and mats, and perhaps also for weaving cloth. Thin branches of fibrous plants such as willows were handy for making baskets.

Not surprisingly, there is no record of the earliest use of fibers. Plant materials are very perishable. They decompose by the action of water, temperature, and fungi and are subject to attack by insects. The oldest residues of plant fibers come from dry areas such as Egypt and coastal Peru and wet ground such as peat bogs. Lack of moisture in the first case, of oxygen in the second, contributed to their preservation. Contact with metal has preserved other ancient fibers, such as cords for binding spearheads. We know how some fibers such as mats and baskets were used because of impressions on ceramics. In other cases, evidence is supplied by pictures or artifacts such as spindle whorls, which are left when fibers vanish.

Plant fibers that people use come from a great variety of species and tissues. All higher plants have fibers. Tree stems are the main source of fibers. Wood fibers, the tissue that gives wood its elasticity, are stiff. The quantity and quality of fiber give different woods their unique characteristics. Wood fibers are hard to separate from the rest of the tissues of the stem, although that process is part of making paper. Fibers also come from the herbaceous stems of species such as flax, hemp, jute, and ramie. Some leaves, for example, manila hemp (a species of banana) and Mexican henequen, are a source of fiber. Fibers from herbaceous stems and leaves are more pliable than wood fibers and can be separated more easily from the rest of the tissues. Finally, an important source of fibers is seed hairs, especially cotton.

Flax is one of the oldest cultivated plants, having been domesticated in the Fertile Crescent some 9,000 years ago. Originally it may have been grown for food, since the seeds are edible and rich in oil. Today, it is grown primarily for linseed oil, the base for most oil paints.

Linen fibers are the stem or "bast" fibers of flax. They consist of thin, elongated cells with thick cellulose walls. Linen fibers can be as long as one meter. Plant breeders have produced special fiber varieties with long stems, in contrast to the oil varieties that are much shorter but produce many more seeds. Linen is probably the oldest plant fiber. Archaeologists have recovered fragments of linen cloth from Swiss lake villages dated at 4500 BP and from ancient Egypt dating back over 6,500 years.

Egyptians specialized in the weaving of fine linen cloth, which they exported all over the Mediterranean. The oldest remaining cloth, which is some 7,000 years old, has 20 to 30 warp threads per inch. (For comparison, Harris tweed has about 24 warp threads per inch, cotton bedsheets usually have 180 to 200 threads per inch, and most modern clothing fabrics are somewhere in between.) This ancient specimen is good-quality cloth and definitely beyond the experimental phase, so weaving must have been around for a long time by then. Finer linen remnants from about 5000 BP have 88 warp threads per inch. By the First Dynasty (4800 BP) Egyptians were weaving very fine linen with 160 warp and 120 weft threads per inch. The Egyptians also learned to dye linen and produced red, yellow, green, and blue fabric. They made linen into fine clothing and impregnated it with resin to wrap mummies. The Egyptians believed that wool was unclean and did not use it much.

Palestine was also a center of linen production in ancient and classical times.[7] In Proverbs 31:13 a virtuous woman "seeketh wool, and flax, and worketh willingly with her hands." In Europe the earliest traces of flax come from a wild species. Archaeologists have found remains in sites on the Danube in southeastern Europe dated at 7500 BP and in Swiss lake villages from 6200 BP.[8] Sites from the late Neolithic in Switzerland suggest a transition to the cultivated form, which presumably came from the Middle East.

Plant fibers need a fair amount of elaboration before they can be used. The fibers have to be separated from other plant tissues before being spun into yarn. Then they are woven into some kind of fabric. Before humans could replace skins with plant fibers they had to reach a certain cultural milestone in the production and use of tools.

The traditional way of separating fibers such as linen from plant parts is called retting. The parts are immersed in water so that tissues decompose. The fibers, made of cellulose and often impregnated with other substances such as lignin, are more resistant. The weaver dries the wet mass and then either tramples or crushes it with rollers to isolate the fiber from remaining tissue.

Spinning is a way of making a long cord out of short fibers. Some kinds, especially short fibers like cotton, are prepared for spinning by carding, which makes the fibers approximately parallel. The first step in spinning is to take some fibers from a bundle, which may have been carded, and draw them out into a section to be spun. Drawing makes the fibers more parallel and determines the diameter of the yarn. The second step consists of twisting the fibers so that they cling together. Weavers often perform these two steps simultaneously. Only fibers formed by cells with rough surfaces can be spun. Human hair and most

*Spinning with one spindle. From an Egyptian
tomb painting, Beni Hasan, c. 1900 B.C.*

kinds of dog hair are too smooth for spinning; the fibers would not
cling.

Spinning can be done without tools by rolling drawn fibers between
the hands, or between hand and thigh or hand and cheek. Presumably
the first spinners did it this way. The resulting yarn is relatively coarse.
A higher-quality yarn can be made with a spindle, a short tapered stick
usually weighted with a whorl. When the spindle is twirled, so are the
fibers. Some spindles are rolled against the thigh, some are rolled between the hands, and some are twirled and dropped. After the weaver
has spun a section of yarn, he winds it around the spindle. The spinning
wheel, which was probably invented in India, has a spindle twisted by
action of a wheel. Even today, an Indian spinner will sit on the ground
by the wheel and turn it by hand.

Archaeologists have found ancient crude cords made by rough spinning. Their makers used them to tie things together or looped or knot-
ted them into nets. Net making preceded weaving, as did braiding and
twining techniques used in the production of bands and mats.

Basketry employs a variety of techniques, a common one being coil-
ing. The basket maker lays down a foundation of thick fibers, frequently unspun, in a spiral shaped to the desired form. The coils are po-
sitioned and then bound to other adjacent coils by another fiber. The
oldest remnants of baskets are the same age as the oldest fragments of
woven cloth. Many authorities believe that basketry preceded weaving, but the jury is still out.

Mat making was likely weaving's direct predecessor. Mats and cloth

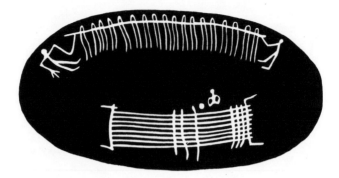

Horizontal loom. This is the earliest representation of a loom, taken from a tomb in Badari, Egypt, c. 4400 B.C.

are made by interlacing two perpendicular sets of fibers. We call the lengthwise set the warp, the perpendicular set the weft. True weaving, by most definitions, requires a loom. A loom is a device that stretches the warp fibers so they can be more easily interlaced with the weft fibers. Loom weaving, like agriculture, appears to have been invented in several different places. From those places it spread to the rest of the world. The reason it is believed that more than one group invented weaving is that different regions of the world adopted different kinds of looms and fibers.

The most primitive kind of loom is a simple frame loom in which four sticks are lashed together to form a square or rectangle. The weaver strings warp threads from one end to the other, then interweaves the weft threads. A related loom uses a piece of wood shaped like a bow, with many warp threads instead of a bowstring. This loom lacks a heddle, by some definitions essential to a true loom. A heddle is a device that lifts a group of warp threads all at once so that the weft thread can be passed from one side to the other in one motion.

One of the looms with a heddle is the backstrap loom. The weaver stretches the warp between a fixed point, such as a tree, and a strap that goes around the weaver's waist. The heddle is usually a bar with string looped around it and around the threads to be lifted. The weaver controls the tension on the warp by leaning back. This loom seems to have

Backstrap loom. From a pre-Columbian vase from Peru.

originated in Asia because it is known from the East Indies to Korea. It is also known in the Americas. The backstrap loom made possible some of the world's most extraordinary textiles, those of the pre-Columbian Andes. It is still in use in parts of Mexico and Central and South America.

Another kind is the horizontal loom. Ancient Egyptian pictures and models show looms of this type. To hold the warp, the horizontal loom has two bars fastened to pegs or trees. Nomads could easily dismantle and transport this type of loom. The vertical loom was probably invented in the Middle East. It is similar in principle to the horizontal loom but is suspended from a tree or roof beams. The ancient Egyptians used it beginning in 5500 BP. Some weavers still use a version of it for tapestry and rug weaving.

Still another type of primitive loom is warp weighted and vertical. The weaver ties stones or pottery weights to groups of warp threads to provide tension. Weavers from the Middle East north to Europe used this type of loom. Warp weights dating from 5000 BP have been recovered from Troy. They are the oldest relics of looms in existence, though there are earlier pictorial representations.

The oldest woven cloths known are from the Middle East. From Egypt we have remnants of linen cloth dated 7000 to 6500 BP. There

Odysseus and Circe with a warp-weighted loom. From a Greek vase,
fourth century B.C.

are also traces of ramie from Egypt dated 5000 BP and cotton from India
(5000 BP) and Peru (3500 BP). Animal fibers, wool and silk, have also
been found but are not as old as plant fibers because they are more vulnerable to insect attack. The Chinese domesticated the silkworm at
least 7,000 years ago.

Weaving is an example of a craft that flourished with the development of cities. Freed from having to raise or collect their own food,
some people learned and improved on the specialized skills needed to
prepare fibers, make looms and other tools, and weave cloth. Like agriculture itself, weaving evolved gradually.

Spinning and weaving and all the other manifestations of specialized
labor in early civilizations would not have been possible without successful agriculture, and successful agriculture depended on irrigation.
Learning to work with water was a landmark in the transformation of
the human habitat. In desert and semidesert areas agricultural yields
were low and unpredictable if the only source of water was rain. Permanent rivers filled this void, supplementing rainwater or replacing it
in times of drought. But if irrigation made agriculture in areas such as
Egypt and Iraq possible, it created a new set of dependencies and environmental problems. River water, unlike rainwater, which is almost
pure, carries silt and salt in suspension. When river water was used to
irrigate fields, deposits accumulated. Silt could fertilize a field, but salt

usually impeded the growth of plants. In many parts of the world so much salt accumulated in irrigated fields that they had to be abandoned. Thus irrigation often acted as the destroyer of its own success, a phenomenon that has resurfaced throughout history as humans have tried to direct nature toward their own benefit.

CHAPTER

Agriculture Spreads to Europe

BEFORE PEOPLE COULD adopt agriculture as the primary method of obtaining food in what are now the dryland farming areas of Europe, they had to learn new agricultural techniques. The soils of southern Europe were not sufficiently rich or plentiful to sustain civilization based on primitive swidden agriculture, and there were no large river valleys comparable to that of the Euphrates or the Nile to sustain a civilization based on irrigation. The problem was how to keep the soil producing year after year. The methods they developed—crop rotation, fallows, and animal manure—are still the principal methods employed by farmers to restore fertility today.

When agriculture first arose in the Fertile Crescent, it served as a supplement to hunting and gathering because early agriculture was not very productive. The chief problems were the unreliable rainfall in the Middle East and the loss of fertility in soil planted with crops year after year. Furthermore, primitive tools were inefficient. When, centuries later, people moved to the fertile valleys of the Euphrates, Tigris, and Nile, they developed more intensive cultivation methods that allowed them to rely primarily on planting and herding for their food needs. They discovered that irrigation would give moisture to their plants and that deposits of silt would restore fertility to soil. They invented the plow and harnessed animal power to increase labor productivity.

As farmers found ways to increase agricultural production, human populations expanded. Paradoxically, an increased population soon made the new techniques insufficient. For example, in Babylon irrigation expanded productivity and allowed population density to increase. However, the larger population required a greater cultivated surface and the expansion of the canal system. With time canals became silted, virgin land ran out, and soil was depleted of its nutrients. The history of agriculture traces a repeated pattern of population expansion followed by the discovery or invention of ways of increasing food production, which in turn was followed by another population expansion. Meanwhile each new cultivation method had a greater impact on the landscape than its predecessors did.

We have seen how the invention of the plow and the use of oxen increased the productivity of the Mesopotamian farmer. Early plows were very light, in essence modified hoes. However, like all useful inventions the plow was improved and made more efficient. At some time this technology diffused into the agricultural villages in the hills and mountain valleys that surrounded the Mesopotamian plains.[1] The plow in turn permitted the introduction of the short fallow as a way to improve soil fertility.

The slash-and-burn technique used in early farming (and still in parts of the world today) requires more land than any other kind of farming. The farmer leaves the land fallow for a relatively long time, usually more than ten years. In that time nutrients from litter and rainfall, especially nitrogen, are incorporated into the soil, then passed to the natural vegetation. When the farmer burns the vegetation to prepare the field for planting, the soil reabsorbs some of these nutrients in the form of ash. A great proportion of the nutrients, volatilized by fire, are lost. Eventually they dissolve into rainwater and are incorporated back into the soil, but in the short run they are lost to the slash-and-burn farmer. Because the ash contains only a small portion of the nutrients, the farmer must burn a lot of plant material to obtain sufficient nutrients to nurse a crop. If instead of burning the vegetation the farmer buries it and allows it to decompose, most of the nutrients can be recovered and used to improve soil fertility. Burying woody vegetation using primitive stone tools is almost impossible. But burying herbs, the first species to grow in abandoned agricultural fields, requires less effort.

The adoption of the plow and animal power allowed villagers in nonirrigated lands of the Fertile Crescent to farm their lands more intensively, rotating crops with one or two years of fallow rather than ten or more. The secret lay in not allowing weeds and especially woody vegetation (shrubs and trees) to become too dense. To prevent overgrowth during the fallow period, the field had to be plowed once or even twice a year. Rotation of cereal crops with several years of fallow increased agricultural productivity and made rain-fed agriculture more reliable.

However, short-fallow agriculture created a new problem, namely, soil erosion. Soil on bare fields, especially in hilly terrain, easily washed or blew away. The destruction of oak and pistachio forests in the mountains of the Fertile Crescent resulted in enormous losses of soil and made much of the region barren and unproductive.

Although European people domesticated and introduced into cultivation several crop species, most notably oats, they did not invent agriculture.[2] The techniques of planting domesticated species reached Europe from western Asia. Two aspects of this diffusion are intriguing. First, it took almost 3,000 years for wheat and barley agriculture to spread from Asia Minor to Greece, a distance of barely two hundred miles. Second, it took another 3,000 years for agriculture to predominate in Scandinavia and the British Isles. This tells us that the hunter-

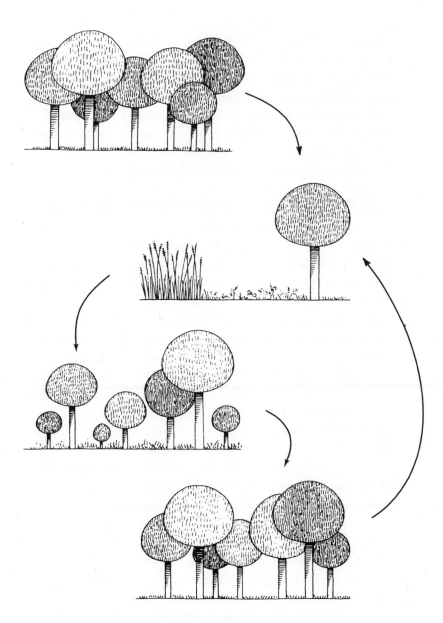

Outline sketch of slash-and-burn agriculture. After cutting and burning a clearing in the forest, farmers cultivate for one to three years, then abandon their fields when productivity is reduced. After ten to twenty years, the soil recovers its nutrients and the field is ready for a new cycle.

gatherers who occupied those areas were satisfied with their methods of obtaining food. And for early planters agriculture was not the only way of procuring sustenance. It coexisted for a long time with hunting and gathering, especially in the northern part of Europe. Furthermore, people might not have always made the change from hunting and gathering to agriculture voluntarily. Invaders interested in benefiting from agricultural surpluses could conceivably have forced people to change their style of life.[3]

The first evidence in Europe of a well-defined cultural complex of agricultural sedentary people is from the island of Crete and in mainland Greece, Bulgaria, and southern Yugoslavia. Traces of these people show up abruptly in the archaeological record and are well established by 7000 BP. They were communities of farmers who cultivated emmer wheat, two-rowed barley, and lentils, herded sheep and goats, knew the plow, and practiced short-fallow rotation. The pattern suggests colonization by farming people from the east. In southwestern Europe—present-day southern Italy, southern France, and Spain—the earliest relics of agriculture date from 7000 BP. There are indications from southern France that the domestication of animals preceded agriculture by a considerable time. In central Europe the first evidence of agriculture based on the cultivation of emmer wheat is much later, from the Bandkeramik culture of about 6000 BP. In northern Europe— northern Germany and Poland, Scandinavia, and Great Britain— evidence for agriculture dates from about 5500 BP.[4]

During the fifth millennium BP there were general and dramatic cultural changes throughout the Mediterranean area. In the Aegean this period witnessed the key transition from the Neolithic to early bronze culture. The number and size of agricultural villages grew, wealth increased, and societies became more hierarchical.[5] These developments coincided with the introduction of olive and grape cultivation. Furthermore, the cereals and pulses that early farmers grew together were now separated, with the pulses close to the settlements and the cereals further afield.[6] There is also clear evidence of increasing population. All indications are that these agricultural societies had moved away from growing crops only to cover their needs to producing a surplus. Similar developments would take place a little later in Italy, southern France, and Spain.

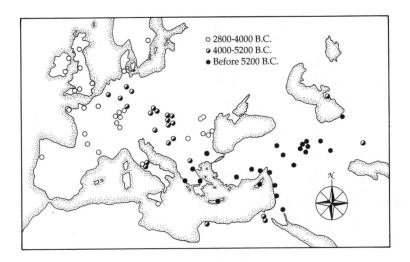

Archaeological records of agriculture in Europe. Note how all records
before 5200 B.C. are from Asia Minor or the lower Balkan peninsula.
The earliest records in central Europe come from between 5200 B.C. and
4000 B.C., and those in western and northern Europe date from between
4000 B.C. and 2800 B.C.

The first people to colonize Crete arrived in the ninth millennium BP,
presumably from Asia Minor. Further immigration took place at intervals, also from Asia Minor, as attested by excavated pottery. Within an
area of 160 by 30 miles, Crete provided enough land for its early inhabitants to grow food. Early Cretan farmers cultivated emmer and einkorn wheat, barley, peas, lentils, vetch, linseed, figs, olives, and grape
vines. At Knossos the Cretans cultivated the more advanced hexaploid bread wheats. Grapes, originally from the Caucasus and diffused
by way of Mesopotamia to the Mediterranean, were an important crop.
Farmers irrigated the grape vines and the olives, originally imported
from Asia Minor and the major agricultural export of the islands, with
spring water. Archaeological excavations at Knossos and Myrtos have
unearthed presses for wine and oil as well as large storage vessels. Cretans kept sheep, goats, pigs, and cattle in that order of importance.

In the fourth millennium BP the Minoans on the island of Crete became the first major civilization to emerge in Europe. Minoan society
was intricately hierarchical, with classes of dominant leaders, bureau-
cratic officials, craftsmen, merchants, scribes, priests, and agricultural

workers. Minoan society produced remarkably beautiful art objects in pottery, stone, and metal. The Minoans also built large and elaborate structures referred to as palaces by modern archaeologists.[7] In certain areas human density was high. The population of Knossos has been estimated to have been 100,000 in the fifteenth century B.C.[8]

In classical mainland Greece farming was limited, as it is today, to small isolated plains and valleys. Of the approximately 300,000 hectares of land in Attica, only a third was suitable for agriculture. Although the Greeks used the plow, it was a light scratch plow that did not turn the soil but rather opened furrows. Consequently farming was restricted to alluvial and loess soils. Alluvial soil is formed by sediments deposited on river floodplains. Loess comes from wind-carried material. Both types tend to be free of rocks and stones and are easy to work, especially with primitive scratch plows.

Early Greek farms were mostly small, independently owned plots. Only citizens could own land. The livelihood of the Greek city-state depended on the produce of these small lots. Cereals, peas, lentils, olives, figs, and grapes were the main crops in ancient Greece. Cereal production sufficed only to cover the needs of one-quarter to one-half of the population. The remainder was imported, traded primarily for olive oil and ceramics. The olive occupied a central role in the Greek economy as a source of food, cooking oil, illumination, and fuel.

Little is known about the population of ancient Greece. According to a census taken by Demitrius of Phaleran in the fourth century B.C., "there were 21,000 citizens, 10,000 metics [possibly foreigners and not much better off than slaves] and 40,000 slaves in Athens."[9] By then farmers and stockmen occupied all the habitable area of Greece, and the population may have been about 2.5 million.

The diet of the ancient Greek was mainly vegetarian. It consisted of a simple fare of porridge, bean broth, pulses, olive oil, fish, oysters and other shellfish, goat's milk, cheese, wine, and bread made mostly of barley. In Homer's day people were called grain eaters because they consumed a wheat and barley porridge called *sitos*. This was made from grain ground in a mortar and served with salt and honey. Barley was also made into flat griddle cakes, or *maza*. Leavened wheat bread called *artos* was a luxury in the early days of classical Greece, available for the

Olive harvest in Greece with the use of sticks. This harvesting system is still in use. Taken from a black-figured vase of the sixth century B.C.

most part only to the rich and served on feast days. Because the barley cakes were not very flavorful people often dipped them in honey or vinegar.[10] Honey from the mountains of Hymeltus near Athens was famous throughout ancient Greece. The Greeks abhorred cow's milk, which they considered unwholesome. Cheese was made primarily from goat or sheep milk.

With the expansion of Greek civilization and the growth of trade, the diet changed in the course of the sixth to fifth centuries B.C. Wheat bread and porridge became more prominent, as did fish. Barley bread was now relegated to the poor. Spices such as cumin, coriander, sesame, and silphion entered the diet, and sauces were more common. A famous one, called *garon*, was prepared by leaving fish immersed in salt brine to ferment for two to three months. A diversity of vegetables appeared in the menu, including asparagus, cabbage, cucumbers, spinach, and celery. Cooking grew into an art, and there is mention of cooking treatises, famous chefs, and the organization of dinners for special occasions.[11]

Wine was central to the Greek diet and culture. It had religious and social connotations and was widely celebrated by the poets. "When

men drink wine they are rich," said Aristophanes. According to Eurip-
ides, "Where there is no wine, love perishes, and everything else that
is pleasant to man." It is likely that the Greeks diluted their wine. They
made it from sun-ripened grapes heavy in sugar. One type of popular
wine was made from the syrupy juice that dripped from overripe grapes.
The Greeks even made wine from raisins. They used clay storage vessels lined with resin that gave the wine a strong flavor similar to that of
present-day Greek retsina wines.

Greek farmers produced more than they needed of certain foods
(primarily olives, olive oil, and wine) but not enough cereals. Thus,
many city-states could not produce enough food to feed their citizens.
They extended their food supply by commerce or plunder and by the
colonization of richer farmlands in the Mediterranean, especially
southern Italy and on the shores of the Black Sea, southern Russia, and
the lower Danube plain. The Greeks made little distinction between
commerce and piracy. Which they chose depended largely on the military strength of the party they came in contact with.

The Greek colonies specialized in the growing of cereals and the
smelting of metals. They conducted a brisk trade with the homeland.
At the mouth of the Rhone River, Greek colonists from Phocaea on
the Asia Minor coast founded the town of Massalia (present-day Marseilles). It conducted a brisk trade with the tribes of interior Gaul. In
exchange for cereals, amber, and tin, the Greeks provided Gaul with
amphorae of wine. Archaeologists have found remains of amphorae in
the interior of France as far as the valley of the Seine.[12] Toward the east,
Greek colonies in the Crimea, southern Russia, and the lower Danube
produced large quantities of cereal. These, together with furs, timber,
and slaves, the Greeks exchanged for wine, olive oil, and manufactured goods such as ceramics and precious objects. The extensive production of grain in the Greek colonies allowed farmers in mainland
Greece to further concentrate on the growing of olives and grapes.

A Greek creation that greatly simplified foreign trade was the coin.
Before the advent of coins all commerce was done through barter. The
Babylonians and Egyptians used money in a limited way, purely to assign value to commodities. Exchanges involved merchandise on both
sides. With the invention of coins (attributed variously to King Midas

or Croesus), traders could sell goods for silver coins, which became a commodity in and of itself, but one with particular characteristics.[13]

People had traded metals, especially precious metals, before. Their weight and purity had to be confirmed in each transaction. The invention of coins, which carried the stamp of the issuer and an agreed value, greatly simplified commerce. The names of the early coins, including drachmas and shekels, referred to units of weight, because a coin represented a given weight of metal. Early coins often bore the stamp of an ear of barley or wheat, a reminder that originally food (probably barley) was used as a unit of exchange.[14]

The Greeks learned that control of land and agricultural surplus were the keys to wealth and power. Through both conquering and trade, they helped diffuse knowledge of their crops and animals. It is likely, however, that for the most part their crops preceded them. Wheat, barley, beans, lentils, olives, and grapes slowly occupied all the Mediterranean basin and then northern Europe until climatic factors checked their progress. The early agricultural areas of the eastern Mediterranean, especially Egypt, Anatolia, the southern Russian plain, and the Levant, continued for a long time to be areas of surplus production. They exported grain to Greece and later Rome, which, as opposed to the rest of Italy, was chronically short of it.

In the third and second centuries B.C. Greece was plagued by wars and revolutions that led to famine and poverty. Because people were so dependent on the produce of their fields, the Greeks perfected a form of warfare that had its origin in the Middle East, namely, destruction of the enemy's crops.[15] During the Peloponnesian War, clashing Greek armies repeatedly destroyed the olive groves on which the economy rested. Olive trees take fifteen to twenty years to produce for the first time, forty years before they are in full production. Their loss would cripple the rural economy of Greece.

Farming in Greece coexisted with hunting and gathering well into Roman times. Hunter-gatherer tribes retreated into more mountainous areas while farming was practiced in the alluvial and loessial soils of river valleys and outside cities. At the beginning, Greek farmers must have followed the swidden agriculture first practiced in the Levant of planting a field until exhaustion, then moving on to another. But this

practice required a lot of land.[16] Because of their small plots and in-
ability to expand, Greek farmers eventually had to develop ways of restoring fertility to land already under cultivation. The pattern has
been repeated throughout history all over the world.[17] At the beginning there is ample land and farmers practice some kind of swidden agriculture. As population increases, the rotation time gets shorter until
eventually farmers need to cultivate all the land permanently. To
maintain acceptable yields, ways of restoring soil fertility must be
found. The three methods developed in classical times were crop rotation with fallowing, fertilization with animal manure, and the inclu-
sion of a leguminous species in the rotation.

2,500-year-old Greek coin with a figure of an ear of six-rowed barley.

Rotation of crops with fallow consists in its simplest form of alternating a year of crop with a year of plowed but not sown field. Initially
the Greeks plowed fallow fields only once, but later they plowed them
twice or three times. The decomposition of weedy, plowed vegetation
added organic matter to the soil. Once plowed under, weeds could not
flower and seed that year. The fallow also reduced harmful fungi and
other pests, evaporation from soil surfaces, and transpiration by plants.
The increase in nutrients came in part from rainwater, in part from the
activity of free-living, nitrogen-fixing soil organisms.

Animal dung and urine contain most of the nitrogen and phosphorus from the fodder livestock consume. By pasturing animals in nonagricultural areas and transporting dung to the field, Greek farmers transferred nutrients to the field. In *The Odyssey*, Homer mentioned the importance of manuring vineyards. Theophrastus recommended that manure be applied abundantly on poor soil but sparingly on rich. Manure was so valued in Athens that the city developed a complex disposal system to carry its sewage to the farms and vineyards of the adjacent coastal plain.

The fertilizing effect of leguminous plants was discovered in classical times. Nitrogen is a nutrient essential to all living things. Although it is the most common element in air, it is often scarce in soil. In air it exists in molecular form, which cannot be used by most plants and animals. Some bacteria and blue-green algae can "fix" nitrogen, that is, convert it to a form that higher organisms can use. Most species of the plant family Leguminosae (the family of pulses) form a special symbiotic association with nitrogen-fixing bacteria. Herbage of these species is rich in nitrogen, and when it decomposes after being plowed under it acts as a fertilizer.

The climate in Greece is typically Mediterranean, with winter rain and summer drought. Summers can be hot, with average temperatures above 30° C (90° F). Farmers sowed cereal seeds by broadcasting them on plowed fallow fields in the fall. In February or March they hoed the fields to remove weeds, and the crop was ready for harvesting with sickles in late May or June, before the fields dried out. Domestic animals, mostly sheep, were pastured on the stubble. Farmers plowed fallow fields two or three times a year. The first plowing took place in the spring. The fields were crossplowed after the harvesting of the summer crop. The third plowing, in fall, followed the direction of the first in preparation for sowing. Greek plows were lightly built. The beam and the share were usually made out of oak wood, the pole out of laurel or elm. Greek plows penetrated the soil only about ten centimeters. Farmers removed weeds with the help of hoes. They threshed the grain with sticks or the hooves of oxen.

Roman agriculture was similar to Greek. In the early days of Rome the Italian countryside characteristically consisted of privately held rural plots of a few acres worked by the owners. Cereals and legumes alter-

Greeks plowing with primitive scratch plows. These consist of a sharp, pointed hardwood share and handle pulled by an ox. From a black-figured Nikostenes cup, sixth century B.C.

nated in a two-field rotation that survived until modern times. Oxen served as beasts of burden and as sources of fertilizer and food. The culture of the fig, the olive, and the grape vine slowly encroached on cereal agriculture. Yields were low and farming methods still primitive, cultivation was with plow, hoe, and spade, and many of these small farmers were in debt to patrician city dwellers.[18]

The principal food of the Roman was a thick cereal soup, bread, and wine, supplemented by turnips, olives, beans, figs, goat's milk, and cheese. Beans, peas, turnips, cabbages, and onions were the principal vegetables, and apples, plums, pears, cherries, apricots, and pomegranates important fruits. Spices such as garlic and herbs served as condiments. Pork was the main meat. Those who lived close to the sea had access to fish.

Cereals, primarily wheat and oats, were pounded in a mortar, mixed with water, and cooked into a porridge called *puls*. For a long time this dish was considered the national dish, rather like oatmeal porridge in Scotland. *Puls* could also include lentils or beans. Our word *pulse*, for edible leguminous seeds such as beans, peas, and lentils, comes from *puls*. Mills for grinding grain gradually came into being. An excavated Pompeian bakery contained several hand-turned mills with which the baker prepared his flour. The traditional Roman bread, *libum*, was flat like a pancake and cooked in hot embers and ashes. Romans continued to consume it even after the advent of leavened bread. Although Romans were aware that whole-wheat bread was more nutritious, like so many people today they preferred white bread.[19] Gladiators and slaves ate barley bread, as Romans attributed to barley special invigorating characteristics.

As time went by, many small farmers left the land to participate in Rome's wars. Burdened by debt and enticed by the city, many smallholders sold their land to wealthy urban families. Apparently patricians also encroached on public lands. Increasingly, large holdings (*latifundia*) belonging to absentee landowners and worked by enslaved prisoners of war replaced the small farm as the characteristic rural enterprise.

The typical rural settlement at the beginning of the Roman empire was the *villa rustica*, a complex of fields, houses, and farm buildings. Normally such large enterprises combined agriculture with stock raising. Primarily they occupied the lowland areas, although they could also include drier upland areas suited only for stock raising. The owners typically lived in a nearby city but closely supervised operations.[20] A *villicus*, or slave steward, oversaw the laborers. Besides slave laborers there were also free tenant farmers and renters who worked on the estates, and some smallholders who tended their own land. In later imperial times land became concentrated in fewer and fewer hands. Many rich owners built elaborate and luxurious buildings, the remains of many of which survive.

These estates grew many crops. "Of all fields the vineyard is the best," wrote Cato, "if it produces plenty of good vines; in the second place is a garden that can be watered; in the third place, a willow grove; in the fourth, an olive field; in the fifth, a meadow; in the sixth, a corn [wheat] field; in the seventh, a wood that grows up again after it is cut; in the eighth, a field planted with trees for vines; and in the ninth, a wood for masts."[21] The basic staples were the cereals, primarily wheat and barley but also oats and millet. Proso, or true or common millet

(*Panicum miliaceum*), is among the oldest and hardiest of crops. Called *milium* by the Romans, it is native to the Mediterranean region and was cultivated there until modern times. The dehusked grains were boiled and cooked like rice or ground and made into porridge. The Romans also grew legumes such as broad beans, lentils, and vetch, and recognized their importance in restoring soil fertility. In the words of Virgil,

> Where vetches, pulse and tares have stood
> And stalks of lupines grew (a stubborn wood)
> Th' ensuing season, in return may bear
> The bearded product of the golden year.

Olive, fig, and especially grape cultivation grew in importance. Near cities Romans cultivated a large diversity of flowers and vegetables such as radishes, turnips, lupines, spinach, and artichokes. (The artichoke probably derives from the thistle or cardoon, cultivated for its edible stalks and leaf petioles. What we eat in the artichoke are the fleshy bracts surrounding the head, or inflorescence.) Different species of fruits and nuts were common: cherries, peaches, plums, quinces, walnuts, chestnuts, and almonds.

The most important animals were sheep for wool, cattle for labor,

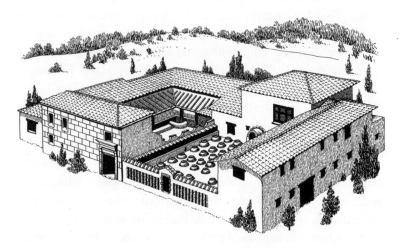

Model of a villa rustica winery from central Italy from the last years of the Republic.

manure, and meat, horses for warfare, and bees for honey, the chief sweetener in those days. Although the Romans grew flax, linen was not as important as wool cloth.

The Roman knowledge of agronomy, well developed, has come to us through the writings of Cato, Columella, Pliny, Palladius, Seneca, and others. The Romans were discerning observers, acutely aware of the importance of soil, climate, and proper care for success in agriculture. The Mediterranean climate of mild, rainy winters and hot, dry summers meant that most crops had to be raised during the winter and spring months. There were two seasons for sowing grain, fall and spring. Fall sowing took place between the fall equinox and the winter solstice, usually from early October to the end of November. Spring sowing started when the land was ready to plant and finished in March. Farmers preferred to plant in the fall. According to Cato, "The trimestrian [spring] seed-time may be used where it is very inconvenient to use the early one, and where the land is so rich, that it is fit for carrying crops every year successfully." Or Palladius: "The trimestrian sowing agrees very well with cold and snowy places, where the quality of summer is wet." That is, where the typical Mediterranean climate did not prevail. There were also warnings not to plant in the fall before the rains had soaked the ground, or when the ground was too wet, as seed would suffer from both too much rain and too little.[22] Palladius commented that "lands when sown ought to be in a temperate situation, neither too wet nor too dry; yet if the drought continues for too long, seed harrowed in it is better preserved in the fields than in the granary." From this we may deduce that even if the rains did not come in time, farmers sowed the dry fields in the hope of a late rain.

Wheat and barley ripen in early summer. Farmers used hooks, sickles, or knives to cut off the ears at the base or halfway down the stalk. They normally did not tie the ears into sheaves but took them directly to the barn for threshing. According to Columella, "when corn is ready it must be quickly reaped, before it is scorched by the heats of summer, which are very great at the rising of the dog-star; first, because it becomes a prey to birds, and other animals, and then, because the grain, and even the ears, fall from the parched stalks; and if there should be storms or whirlwinds, the greater part is driven to the ground." Some farming problems have not changed with the passage of time.

The grain was threshed in a round, level enclosure known as an area. The floor of the area was carefully smoothed and mixed with amurica, the watery fluid recovered after oil is extracted from olives, to harden it and keep weeds from growing. The center of the floor was raised. The Romans threshed—that is, separated harvested plants into the straw (stems), grain (seeds), and chaff (glumes and other small particles)— by driving oxen or a heavy plank over the grain. They also used flails. Afterward they winnowed the grain, tossing it into the wind to separate out the chaff.

Farms produced for themselves and exported surplus to the cities. Given the high cost of transportation, proximity to cities was important. Wine and oil were two products that could absorb the cost of transportation and still leave a profit.

Alcoholic beverages made out of grain and fruit were probably the first drugs in Western culture. Evidence of their manufacture 6,000 years ago is found in Mesopotamia and Egypt. Ancient Egyptians had commercial breweries and produced beer in industrial quantities.

Alcohol is not a natural plant product. It is a waste product manufactured by yeast as it grows. Yeast is a microscopic organism that uses any kind of sugar source as food in a process known as fermentation. It is very common and will grow on any sweet substance, even ground-up seeds (which always contain some sugar) and fruit juices. Beer and wine are produced when yeast ferments, respectively, grain and fruit juice.

The grape is one of the world's most important plants. Grapes can be eaten fresh or dried (as raisins), but they are used principally for wine. They grow wild in eastern Europe, the Middle East, and northern India, and were probably domesticated in the Caucasus. From there they were taken to Mesopotamia in prehistoric times. According to one legend, King Jamshid of Sumeria was fond of grapes. His household kept them in earthenware jars, but to his misfortune some of them spoiled. These were set aside. A lady of the court tasted the fermented juice, liked it, drank more, and curled on her pallet for a nap. Seeing this, the

king and then his entire court did the same and thus was wine introduced to Sumeria.[23]

From Mesopotamia the grape vine diffused to the Mediterranean.
Ancient Egyptian inscriptions indicate that the grape was grown there
in 2375 B.C. Murals on Egyptian tombs carry depictions of grape vines,
grape harvesting, and trampling to obtain the juice. The Old Testa-
ment abounds in references to grapes and wine.

By way of classical Greece and later Rome, the grape vine expanded
into all the lands of Europe and north Africa where it will grow. Herodotus mentioned Greece's important commerce in wine, especially with
Egypt. The Greeks cultivated the grape vine wherever they went: Italy,
north Africa, southern France, and southern Spain, including the
best-known wine regions of the world, Bordeaux and Burgundy. The
Greeks called Italy the Land of Vines.

As we have seen, the grape was the most important commercial crop
in Roman times. Romans grew different varieties and appreciated their
diversity. Grapes were grown on trees and trellises, preferentially on
terraced hills and banks surrounding river valleys. Romans aged wine
in barrels, which they invented. Before then it had been kept in earthenware amphorae.

Grape juice changes naturally into wine when sugar-fermenting
yeast is present. When grapes are crushed with their skins, yeast comes
in contact with the juice. It grows in the juice, using the sugar as its
source of energy, and in the process transforms the sugar into alcohol
and carbon dioxide. Alcohol in high concentrations is toxic, and
when it reaches a certain level (between 12 and 14 percent), most
strains of yeast cannot grow any more. The wine is now ready. It is
young wine, and tastes very much like grape juice with alcohol. When
the first fermentation is completed (sometimes even before), and especially if the wine is in a warm place and exposed to the air, special
bacteria start growing on the alcohol and transforming it into acetic
acid. The wine spoils and turns to vinegar.

Since remotest antiquity, the principal problem was not to make
wine but to keep it from spoiling. In classical Greece (and possibly also
in Egypt and Mesopotamia) wines were drunk young, and most were
probably vinegary. To counter the acid taste a variety of methods were

employed. Greeks learned that air speeds up the spoiling process. Amphorae had narrow necks to reduce the contact of wine with air, and they were kept tightly stoppered. Because air could penetrate earthenware surfaces, the Greeks lined their amphorae with resin. This preserved wine for use in commerce. The Romans took a step forward when they invented wooden barrels. They could be stoppered better than amphorae, and the oak imparted flavor to the wine. Some Roman wines apparently kept very well. There are reports of vintages that lasted up to a hundred years.

The original vegetation of the Mediterranean was dry woodland replete with oaks, such as the cork and the holm oak, and pines. There also grew a large number of smaller trees and shrubs, for example, wild olives, myrtles, rosemarys, lavenders, citruses, and tall heaths. The transformation of vegetation started with the first agricultural settlements of Greeks and Phoenicians in southern Italy and Sicily. The Greeks farmed the alluvial plains, and while they probably did not transform the Italian landscape extensively, they set in motion a process of erosion that continues to this day.

The Etruscans occupied the coastal plains in the west of Italy north of the Tiber River. Part of this area was swampy, and the Etruscans and later the Romans built extensive drainage works. Rome was surrounded by a broad, marshy plain irrigated by the Tiber River and known as Latium. This was drained and transformed into a productive region in the days of the Roman Republic. In an area of about 400 square miles lived between 200,000 and 400,000 people. To feed them all the land was intensively farmed, leading to erosion and loss of fertility.[24] When Rome initiated its many wars of conquest, the drainage works fell into disrepair and the land reverted to swamp once more.

The expansion of Rome and Roman agriculture in Italy created extensive landscape transformation. Hillsides were stripped of their trees, which were used for construction and fuel. Olives and grape vines replaced them. This exposed the bare soil to wind and water erosion. Additional damage was done by grazing sheep. Thus the Romans continued the process of land degradation that began in classical Greek times, continued until recently, and accounts for the desert and semidesert conditions found throughout much of the Mediterranean basin today. The region has been so disturbed through grazing, wood-cutting, and agriculture that only remnants of original vegetation—scrub, known as *garigue*, which grows in limestone soil, and a denser and taller shrub, called *maqui*, which is found in siliceous soil—remain.[25]

The Medieval Farm

STARTING IN THE fifth century A.D. and continu-
ing for some three hundred years, people from northern and eastern
Europe, the Goths, Lombards, Ulans, and others whom the Romans
called barbarians, repeatedly invaded the Roman empire, plundering
settlements and farms. Destruction in the countryside disrupted the
economy of the cities, which depended on the country for food and
other products. As order deteriorated and marauding bands of robbers
ravaged the countryside, powerful landlords created personal armies to
defend themselves. Independent farmers sought the protection of the
landlords and entire villages bound themselves voluntarily to an estate. Simultaneously, authorities passed laws forbidding the sale of

slaves away from the land. This ushered in the new institution of serf-dom where people were bound to the land they worked. Serfdom ele-vated the status of slaves and reduced that of tenant farmers.[1]

In spite of the loss of city markets, the breakup of the Roman empire did not drastically alter the rural order, at least not at first. The *villa rustica* became the demesne or feudal estate of medieval times, presided over by a feudal lord and worked by tenant farmers, some free, some en-slaved. Next to the manor house there was the village inhabited by serfs, or "villains," who although bound to the lord obtained from him the right to farm special plots of land. The village inhabitants owed al-legiance and work to the feudal lord. Agricultural practices became relatively standardized and in time so entrenched as to be a barrier to progress. To a large extent, the uniformity of medieval Europe's social structure was rooted in invariant agricultural practices.

By the ninth century a new order had emerged. Gradually landown-ers freed the countryside from marauding gangs and economic life re-vived in the cities. By the fourteenth century conditions had improved so markedly that a new crisis, rooted in overpopulation and inefficient agricultural production, developed.

The foundation of feudal society was the fief, a tenement held by a knight in return for military service to his lord. Allodial land, that is, land lying outside feudal property and corresponding roughly to Ro-man private property, became rare. By the twelfth century almost every village belonged in a fief. With time the power of the lords grew until it extended over the uninhabited land that lay between villages.

The agriculture practiced on these lands was not uniform through-out Europe. Soil, climate, and culture determined what crops were grown and the rhythm of farming. Crops were roughly the same as in Roman times, and cultivation continued to be concentrated in allu-vial soil in river valleys or on loessial plains. Uplands were mostly for-ested, and peasants used them for hunting and to pasture animals, es-pecially pigs.

The most widespread system of agricultural production in western Europe was the open-field system, so called because there were no fences or hedges separating plots. Each peasant family obtained a plot of land for its own needs and had to provide some labor (roughly a day a week) to the feudal estate. This took the form of work on the lord's lands for men, and helping in the manor or castle for women. The fields split into several traditional categories, and the uses to which they were put, like the social hierarchies, varied little. There was the permanent meadow, or green, often lying along a brook or river. The spring flood both watered and fertilized the green. From it people obtained fodder for livestock during winter months. Around the green was arable or plowed land, and around the arable land was the common pastureland and the "waste," usually forested. The forest provided fuel and construction materials to the estate and served as pastureland.

The open-field system implied a two- or more often a three-year rotation. Arable acreage was divided into two or three large fields (depending on whether a two- or three-field rotation was practiced). Each field was divided into smaller plots, or "shots," and these were further divided into strips. Villagers owned strips scattered over the fields. This way everybody had access to some good and some bad land, and to different products. In addition, each cottage had an attached garden for growing a diversity of vegetables and fruits.

An important principle of the organization of the open-field system was equality. It was important that peasants of similar station have access to like holdings. The assembled "villains" or council of elders managed the village fields communally, but each farmer took care of his own strips. This traditional system of cultivation left little room for initiative to the individual. Custom firmly established the time and number of plowings, the methods of cultivation, and the order of crop rotation. The estate managed its lands independently but along similar principles. There was a lord's meadow for his animals, a garden for vegetables and fruits, and two or three fields subject to the same system of cultivation as the village lands and planted with the same crops.

The basic unit of cultivation was the amount of land that could be plowed in a day: the acre (England), journée (France), or morgen (Germany). The character of the plow determined the shape of a strip. The most economical shape, which minimized turns, was the long, narrow strip, four rods wide by forty long (roughly 5 by 200 meters, or 15 by 600 feet). Normally several parallel strips formed a unit known as a culture.

Because of the lay of the land, adjacent cultures would often have their strips lying at right angles to each other. The scattering of a farmer's strips over three fields made farming more difficult.

Cereal culture, especially that of wheat, exhausted the land. Farmers, especially in central and northern France, slowly replaced the original two-crop rotation—one year in grain, one in fallow—by a system of three-crop rotation. Under this system, peasants grew winter wheat and barley in one field and spring crops such as wheat, oats, barley, or rye in another, and left the third fallow. This reduced the fallow from half of the surface to a third, but also reduced the wheat surface. A variant of this system was to replace the fallow, either entirely or every other cycle, by a leguminous crop such as peas or alfalfa. The Romans had planted legumes for this purpose, but the change did not become firmly established in northern Europe until the sixteenth century.

In the high Middle Ages, life on the rural estates was hierarchical and unchanging. Although the lord had his own quarters and his own servants, at least once a day he joined with members of the community at dinner.[2] This was an occasion to learn about the important and trivial events of the day, to consult and give directions to the steward, and to listen to petitions and resolve minor disputes. Dispensation of justice was the principal prerogative of the master. Custom established everyone's social position and regulated the pattern of daily life. Peasants were an integral part of a rural community that simultaneously aided and constrained them. The needs of their individual farming units shaped their lives. Orthodox arrangements whose purpose was the successful cultivation of the fields of the village ordained their activities.

In the self-sufficient economies of medieval Europe, reducing the risk of failure was more important than increasing productivity. Loss of the harvest meant starvation. Under such conditions farmers could not afford experimentation. Thus time-tested methods turned into inviolable traditions. In England the system was called the custom of the manor; in France, *assolement obligatoire*; and in Germany, *Flurzwang*. These terms imply the same obligatory round of operations in which each member of the village, serf or free, played his or her assigned role.

The produce of the estate lands provided sustenance to the lord and his court. As the Middle Ages advanced and cities revived, farmers could again sell their rural surplus for cash to the cities. Increasingly, lords accepted cash rents instead of labor from village farmers. The farmers in turn began to assert hereditary rights to the land they farmed (it had been granted originally only for the duration of the farmer's life) and to rent or take over estate lands. This process was most noticeable in western Europe.

The medieval diet, like everything else, was quite uniform. The main food, especially in France, was bread. Leavened bread was made of wheat or rye flour, together or separately, or with some barley or bean flour added. The latter would make the bread more nutritious but also heavier. People preferred wheat bread over rye, but wheat did not grow well in the cold north, where rye replaced it.

Medieval bread was altogether different from today's bread. It was more compact and made from the entire grain (bran, germ, and gluten). Forms of bread varied. White wheat bread baked into rolls was the preferred form. Only the lord and those well off had access to this type of bread. The ordinary farmer ate white bread only in good years. There was also a coarser bread for the well-to-do, baked flat and sliced into two round halves that served as plates. On these bread plates diners placed their individual portions of meat during the meal. After the meal they passed them soaked with meat juices down to the lower tables. Peasants baked coarse bread shaped into large or flat rolls.

A thick soup and beer or ale made from barley formed the meal of the peasant. Peasants cooked this soup in a cauldron, a heavy iron kettle, that stood above the embers or hung from a rafter above the hearth. Into it peasants tossed just about anything edible: rabbits, salt pork, old bread, vegetable greens, turnips, and the remains of previous meals. Peasants rarely cleaned the stock pot. As the pot's contents got low they just added more water and ingredients.[3] Another common constituent of the diet was a cereal porridge made from different types of flour, boiled in water, and served with milk and honey. People ate oats in porridge and used them for animal feed. Millet was a secondary cereal used mainly as forage. Yet because its small seeds kept well, millet was stored as a food of last resort for hard times.

Roasted meat dominated the tables of the upper classes. Beef was scarce. More common were venison and bear, sheep, and goat meat. Peasants ate pork when they had red meat. In areas close to the sea people consumed fish. Cheese, milk, eggs, and poultry extended the diet. Chickens, geese, pigeons, swans, and peacocks were common barnyard fowl. Not only did they provide eggs and meat, but farmers used the feathers for bedding or sold them to make brushes, pens, and arrow fletchings.

Europeans preferred meat to vegetables. Northern Europeans, who had hunted big game longer than other peoples, were particularly fond of their animal food. When supplies of game dwindled as an increasing population encroached on the waste, the rich and powerful restricted access to it. They carved game preserves out of former communal forestland that became the property of the king and the nobility. Farmers had to make do with lesser game, if any at all. As people farmed more land to feed the increasing population, less land was available for this game to feed on and its stocks also dwindled.

From roughly 1150 to 1300, feudal Europe underwent a period of slow economic and demographic growth. With population growth came the need to increase the number of farming units. New farms were carved out of forestland or taken from the meadow and the waste. Some subdivision of existing land also took place. The new properties were not as well suited for agriculture as the old fields, which is why they had not been used before. Furthermore, the reduction of meadow and waste decreased the number of large domestic animals, especially cattle and horses, that farmers could keep. Manure from domestic animals, particularly cattle, was the main source of fertilizer. With less manure, soil fertility and agricultural yields declined. Diminished cattle stocks also meant fewer draft animals, which in turn lowered the productivity of human labor. There being no change in agricultural procedures, population expansion had the effect of reducing average yields, especially those of grain.

Slowly, arable fields, originally confined to valley bottoms, took over most level upland areas. Farmers planted grape vines or fruit trees on hillsides when possible. Forests were cut or modified to allow grazing. The increase in arable land was made possible by the development

*Plow with moldboard from a fourteenth-century English Bible. Note the
donkey pulling the plow together with a pair of oxen.*

of the heavy plow and the replacement of oxen by horses. The invention of the horse collar and the metal horseshoe resulted in an increase
in cultivated surface. Gone in the process were the extensive wood-
lands of early medieval times that teemed with deer, boars, and bears
and shaped so many of the fairytale images of wild and threatening forest that we inherited. By the beginning of the eleventh century an expanding population began to invade the forests, cutting down trees,
cultivating cleared land, and establishing new villages. This movement, especially in France, was akin to the opening of an internal frontier. There arose a group of loggers known as *hospites* who lived in primitive huts in the forest. After clearing land they would leave it to be
settled by others and move to another part of the forest. Lords invited
hospites to their land and offered them special privileges for opening up
the forest. The Church participated in the settlement process by establishing monasteries in remote, unoccupied forest areas.

Enhanced economic activity in the cities brought prosperity.
Within local networks of villages and incipient towns the exchange of
agricultural surpluses for handicrafts increased. A few manufacturing
centers such as London and Brussels expanded and began trading on a
larger scale. This created economic opportunity for the lord of the
manor, who controlled most trade. To increase production, the lord

demanded increased services from serfs bound to the manor. Peasants resisted these demands as being against custom. The lord's assertion of traditional rights to the labor of the farmer meant less time available to farmers for their own fields at a time when they were facing lower yields. The expansion of local markets also brought demands for a different mix of products than those habitually produced in the orthodox estate. For example, in England there was a greater demand for wool and for horses, the leading form of land transportation.

Wine was one of the few agricultural commodities in long-distance commerce during the Middle Ages. One of the most important trade routes was between Bordeaux in Aquitaine and English ports. Ships carried the wine in barrels. The size of a barrel was measured in hogshead units (1 hogshead equaled 63 gallons). One pipe equaled 2 hogsheads and one ton equaled 2 pipes. The capacity of ships involved in the wine trade was measured by the number of ton barrels they could carry, and to this day a ship's displacement is measured in tons.

The apparently staid society of the Middle Ages was never, in fact, completely self-sufficient or invariable. The memory of the Roman empire was very much alive, both in institutions and customs, modified by Christianity. Cities continued to function as markets, albeit reduced, for agricultural commodities. As the insecure political situation resulting from the breakup of the Roman empire stabilized, economic activity increased and so did the number and size of cities. By today's standards, however, they were very few and very small. Unless they were ports, or cultural, religious, or manufacturing centers, most medieval towns were market centers for villages within a radius of about six to twelve miles. Most tradesmen were also part-time farmers. In 1300, London, already the largest town in England, had only 10,000 inhabitants. By 1500, there existed only 154 cities in all Europe with more than 10,000 inhabitants, 65 of which were in the Mediterranean basin. The few large and powerful cities concentrated primarily in the Italian peninsula: Venice, Genoa, Milan, Rome, Florence, and Naples.[4]

Slowly there emerged in the cities a population of artisans who sold or bartered their wares for agricultural products. Trade, in the sense of long-distance commerce, dealt mainly in luxuries and involved mostly the rich and powerful, notably in northern Italy, Flanders, and northern Germany. Agricultural commodities were rarely traded other than locally. Farming remained the economic mainstay and overshadowed any other activity.

The expansion of the medieval rural economy came to a halt in the fourteenth century. Farming, geared toward cereal production and based on the open-field system, could no longer satisfy the needs of the population. The immediate cause was a series of famines and disease epidemics of which the best known is the bubonic plague, or Black Death, epidemic of 1347–53. Disease may have been as devastating as it was because people were already weakened by malnutrition. In any case, population declined abruptly, as did commerce. Prices rose. A further complication was the beginning of the Hundred Years War between England and France, which increased the level of taxation in both countries. The result of all these factors was economic stagnation, rural depopulation, and a reduction of the area under cultivation after roughly A.D. 1350.[5]

The loss in population following the Black Death had a number of effects. Labor costs increased and the profitability of estate farming decreased. The declining prosperity of large agricultural estates after the fourteenth century had different social outcomes depending on location. In western Europe, where population density was highest, many lords could obtain remunerative employment at the court. This reduced the time they devoted to the management of their estates. Since corruption was rampant, any estate not closely supervised by its owners soon became unprofitable. Owing to these difficulties and to the expenses of administration, estate farming yielded good profits only when wages were low and prices high. Owners of unprofitable estates leased or rented them to those who could devote the time and effort to make them profitable, usually the better-off peasants. The result was a decrease in the size of farming units in western and northern Europe.

In central and eastern Europe, there were fewer employment alternatives for the nobility. The depopulation of the rural environment increased the vigilance of estate owners. When peasants abandoned land, lords would seize it. This increased the size of estate lands in eastern Europe—primarily Prussia, Poland, the Baltic countries,

Ukraine, and Russia—at the expense of village lands during the four-teenth and fifteenth centuries.

The agricultural crisis of the fourteenth century and the depression that followed changed the traditional rural attitude. The crisis led to a profound transformation in the manner of producing and exchanging commodities, laying the foundation of today's capitalist mode of pro-duction. A largely local economy based on self-sufficiency and barter evolved into a regional, market-oriented, monetary economy. A co-operative system of production dictated by custom and collective de-cision gave way to specialized farming units, each owner or tenant making individual decisions regarding what, when, and how to plant to maximize profits. Higher productivity became an important aspect of farming, as important, if not more so, than avoiding failure. The changeover was slow and took several centuries to become firmly es-tablished.

The Arabs invaded north Africa and Spain in the ninth century. They introduced to Europe a host of new crops and farming methods. The most consequential crop was sugarcane. Other important crops were rice, sorghum, and hard wheat among the cereals, spinach and eggplant among the vegetables, orange, lemon, coconut, banana, and watermelon among the fruits, and old-world cotton. Most of these plants were domesticated in India, notable exceptions being sorghum and watermelons (African) and wheat (Middle Eastern).

Hard or durum wheats are tetraploid. They differ from emmer in that the glumes are small and the grain threshes free. Hard wheats are so called because of the consistency of their flour, known as semolina, from the Arab *semoules*. They don't make very good bread because they have less gluten, but they are preferred in the preparation of couscous and pasta. In fact, it is likely that the Arabs invented pasta.[6]

Yet the greatest contribution of Arab agriculture to Europe was the introduction of summer irrigation, which intensified cultivation. In European agriculture, with its beginnings in the Mediterranean cli-mate of winter rain and summer drought, the emphasis was on crops that would grow in winter or spring, when the soil had enough mois-ture. All the crops domesticated in the Middle East, such as wheat, barley, and peas, had this growth regime. Also, because soils were poor the method of fallowing developed. Consequently, farmers used their land primarily in winter and spring, and about half of it remained idle.

The Arabs were desert people who in their native lands practiced agriculture only with irrigation. They learned early in their history how to use water to grow crops and to farm land intensively. When the Arabs erupted into the Mediterranean, they brought with them their mastery of water. This allowed them to cultivate land during the dry summer. They also brought crops that were tropical in origin and would not grow well in winter. The marriage of the European winter and spring crops with the mostly Indian and African summer crops allowed a full use of the land. With irrigation, farmers could grow two or even three crops a year in certain areas, such as the Vega of Granada or the valley of the Guadalquivir in Spain. To compensate for the loss of nutrients in soil, the Arabs refined the practice of fertilization, using dung from many animals, ground bones, crop residues, limestone, and ashes, among others. Their influence remained in Mediterranean Europe long after the Spaniards ejected them in 1492.

The population of Europe began to expand again in the 1500s. This brought with it enhanced economic activity that created a heightened demand for agricultural products. Meanwhile the medieval open-field system of cultivation had reached its maximum productivity. What followed was the enclosure of open fields to form the system of individual farms that has persisted to our day. The generalized concentration on cereal farming was replaced by a greater mix of crops that took better advantage of a region's natural characteristics. At the same time, transportation facilities slowly improved, widening the area of influence of city markets and therefore the size of markets reached profitably by individual farms.

The transformation in England has been the most extensively studied. It did not come easily. As is often the case, economic change brought a great deal of dislocation. Poor and landless peasants were the principal victims, and repeated and violent peasant rebellions took place in England and on the continent. The enclosure movement was basically a disentangling of the highly integrated ownership relations of the medieval manor.[7]

The medieval system of cultivation, with the land divided into meadow, arable, and waste, was well suited to an epoch when population density was low and production was geared primarily to self-sufficiency. The division of arable land into two or three units, each cultivated for one crop but with each farmer owning strips throughout, gave everybody equal shares of good and bad land but did not necessarily assure surpluses for sale. As population density increased and city-based markets expanded, the system became increasingly less capable of providing for the needs of the villagers and also producing a salable surplus. As demand for cereals increased, arable land grew at the expense of the waste. The latter was the source of game and timber, and forage for farm animals. A larger population harvesting a smaller waste area led to deforestation and erosion. The increasing needs of cities created strong pressure for specialization of farm production and induced a change from a subsistence to a market-oriented economy. The corollary was a change to individual private management, the core of the enclosure movement.

Several factors contributed to making the traditional open-field system inefficient. The fact that each farmer's portion of the arable was not contiguous became a disadvantage. So, for example, in 1635 in the village of Laxton, county Nottinghamshire, a farmer by the name of John Freeman held 29.6 acres of arable land divided into numerous small plots or strips, some less than an acre, as well as two separate enclosures of 3.4 and 7.3 acres.[8] Having to work so many different plots clearly reduced the efficiency of the farmer. It made the vigilance necessary to ensuring a successful crop much more difficult. Not only were there natural pests and adverse conditions to watch for. Many extant reports tell of farmers encroaching in one way or another on their neighbors' strips. Furthermore, in an open field bad farming practices such as allowing weeds to produce seed, pests to multiply, or domestic animals to run loose affected all the surrounding plots, decreasing everybody's efficiency.

A major reason for the rigid enforcement of common planting dates and crops was that after the harvest cattle and horses were let loose on the stubble to graze. The animals had individual owners but the land did not, so there was no incentive to reduce the size of a herd. Grazing exhausted the fallow field, leading to reduced production.

Enclosure refers to the fencing in of former common or waste land. A hedge or a ditch usually marked the boundary. When a lord initiated enclosure, he usually sought to consolidate the various holdings of the estate—meadow, arable, unimproved pasture, and waste—into a contiguous plot of land with the objective of managing it as a unit. Once this happened, the lord was freed from having to administer his land according to the dictates of tradition. Now he could dedicate the entire estate to one activity. In England, many farms turned exclusively to the raising of sheep, taking advantage of the expanding market for wool.

Enclosure was also initiated by successful village farmers interested in consolidating their scattered plots and strips. The process varied from place to place and time to time. The loser was the landless or near landless peasant who lost ancient rights to the use of the commons and the stubble. In many cases, too, the peasant was deprived of employment in the manor or castle that under the old paternalistic system had been virtually guaranteed. Well-documented cases exist of abuse and outright misappropriation of land by some landlords.

In England enclosure began in the twelfth century but was not widely felt until after the crisis of the fourteenth century. The rapid expansion of the Flemish wool trade after that date produced a strong incentive for the production of wool in the English countryside. Only after the land had been enclosed and withdrawn from cereal culture could sheep be kept all year on the fields. Sheep farming also required less labor. Many marginal tenant farmers were pushed off the land. This trend was reflected in an expression of the time, "Sheep eat people." In the English downs sheep were used to help fertilize the fields. During the day sheep grazed on the chalk downs, which were not suitable for agriculture, and at night, when most of the dung was dropped, they were penned in the fields requiring fertilizer. Farmers also collected dung from animals kept in stalls and spread it in the fields. When animals fed on the stubble of harvested plots, they helped return the nutrients in the straw to the field.

During the sixteenth century the concept of individual landownership took hold. This eroded the medieval notion of personal obligation. In England, the change in the relationship between peasant and lord generally favored the latter. The lords persuaded or coerced peasants into converting their copyhold rights (that is, the right to the product of the land) into leaseholds. Under the new arrangement the tenant had rights only during the period of the lease and no rights whatsoever beyond that. This gave the lord, now transformed into landowner, more flexibility in land use. Subsistence peasant farming was slowly replaced by tenant farming, with some wage employment.

In France the peasant fared better than his English counterpart in the struggle over landownership. Peasant rights to the use of land were transformed into ownership with some residual obligations toward their former lords. Over half of France was demesne land that became the property of the nobility, who also managed to acquire property rights over much former common land. Their control over ovens, mills, and other equipment allowed them to continue charging high fees for their use. Peasant landowners could borrow money from their former lords using the land as security; in times of hardship, when they could not pay back the loan, many lost their property. Nevertheless, on the whole French peasants were much more secure and independent than their English counterparts. This allowed them to resist enclosure and continue to farm collectively. Because the open-field system emphasized subsistence farming and resisted agronomic innovation, it eventually succumbed to problems such as low productivity, poor yields, and frequent famines. These were not resolved until after the French Revolution.

In eastern Europe an enclosure movement of sorts took place. Here the lords increased their holdings at the expense of the peasants, reducing them to serfdom or a rural proletariat. This resulted in the large estates that would characterize Prussia, Poland, and Russia. They specialized in raising cereals. Most of their production was for local consumption, but a small surplus was exported to the cities of the West, especially in the Low Countries. This made up for the reduced cereal production there that resulted from the switch to specialized farming, the movement of peasants to the cities, and lower productivity.

Enclosure and better transportation, as well as the development of financial and commercial institutions, created the potential for a more rational use of the land.[9] Medieval agriculture, being geared to the needs of the manorial community, had to be largely self-sufficient. This explains why every estate had the same kind of rotation, based on grains and complemented by domesticated animals kept on the common pasture and fed in winter months on fodder obtained in the wet meadow. It also explains why there was a need for a waste to provide fuel and the wood for construction materials. The rise of a market-oriented money economy allowed an estate to produce one or a few products efficiently for the market, and to buy from outside the estate those items it no longer provided for itself. By freeing themselves from rigid custom and traditional ways, farmers could become better and more imaginative managers.

Although grain continued to some extent to be produced everywhere, the lands of eastern Europe—Germany, Poland, and the Baltic states—began to specialize in wood and grain production. Spain and England concentrated on wool production, while present-day Holland, Belgium, and northern France slowly became industrial and commercial centers.

The sixteenth and seventeenth centuries saw the initiation of the process by which agriculture changed from a subsistence to a market system. For a long time to come farmers would grow much of what they ate, but in Europe and eventually most of the world the goal of farming became the production of a surplus to create capital.

Friesland from the mid-sixteenth to the mid-eighteenth century led the way.[10] This district in the Netherlands became a specialized dairy farming region in the sixteenth century. From the detailed legal records of the belongings of farmers who died and left minor children as wards of the court, historians have reconstructed the economic and social changes that took place in Friesland at this time.

By the beginning of the sixteenth century, Friesland had already ceased to be a truly self-sufficient medieval economy. During the next two centuries specialization intensified. Some regions increased the arable for grain production, others devoted their efforts to cattle breeding, and most areas concentrated on dairy production. With markets available, farmers could not only specialize in particular kinds of farming but also spend more time on farming and less on activities such as weaving and toolmaking. They could now buy goods and services from craftsmen instead of making everything on the farm.

The purchasing power of the average Frisian farmer doubled over this period. Farmers began to invest their earnings in infrastructure, for example, housing, and their standard of living rose. The typical house up to this period had been a variation on the *oud Friese huits*, one-room buildings that housed people and their domestic animals. The walls were wattle and the roofs were thatch. By the sixteenth century farmers were separating the living space from the stalls and occasionally adding a separate dairy room. In the following two centuries throughout Friesland, farmers replaced these buildings with multiroom brick structures with tiled roofs and a separate barn for the animals. Inside, the simple, massive homemade oak furniture was exchanged for more elaborate furnishings bought from local cabinetmakers. Homespun fabric gave way to industrially made cloth. People replaced iron jewelry with elaborate silver and gold ornaments supplied by city or village-based craftsmen. This was all made possible by the increased productivity and specialization on farms.

From the sixteenth century on, first in western Europe and then more broadly, new agricultural procedures were introduced to increase productivity. First applied in the Netherlands, they spread to England with returning émigrés after the restoration of Charles II. The main improvements involved growing legumes (peas, beans, lentils) or turnips in the stubble of the cereal harvest. These crops were used to feed animals, though they supplemented the human diet in bad harvest years. They increased total food production while raising the level of nitrogen in the soil. Another improvement was the replacement of the rigid three-year rotation—two years of grain followed by one of fallow—with more complex four- and five-year rotations. So, for example, grains might be cultivated for two years, followed by a year of legumes and one or two years of fallow. Furthermore, instead of letting a field lie unplowed during the fallow, farmers planted it with alfalfa or clover. This not only added nitrogen to the soil but allowed more intense grazing, increasing the quantity of manure, which was still the principal fertilizer. The improved soil structure led to significantly higher cereal yields. Total grain production increased in the Netherlands and in England despite a reduction in the area planted. [11]

These trends accelerated in the eighteenth century, preparing the way for the introduction of scientific agriculture in the nineteenth and twentieth centuries. Changes included the replacement of oxen by horses as the main source of animal traction; the improvement of soil management and fertilization procedures with the introduction of chemical fertilizer in the nineteenth century; and the invention of better farm implements—harrows, plows, seed drills, mechanical reapers, threshing machines, and so on. Toward the end of the nineteenth century agriculture in Europe, especially western Europe, became highly mechanized, and by the twentieth century the tractor replaced the horse. Since it took about one hectare of improved grassland to keep one horse, land formerly devoted to grazing was freed for cultivation. Farmers also improved yields by better selection of seeds. The development of genetics gave plant breeding a scientific basis, and it became an important contributor to the tremendous improvement in European agriculture.

Throughout most of this period the population of Europe increased. At the beginning of the sixteenth century more than 80 percent of Europeans were farmers. This did not change until the middle of the nineteenth century, but then the number plummeted. The proportion of farmers today is less than 20 percent, and in some countries such as England it is as low as 2 percent. The majority of the world's population works in industrial and service activities in cities and is fed by a dwindling number of farmers. [12]

Sugarcane and Industrial Agriculture

THE SUGARCANE IS a unique plant. A member of the grass family, it does not look at all like a grass. It is more than three meters tall at maturity, and it is not the seeds as in other grasses that are eaten but rather the contents of the stem, watery, fibrous tissues filled with sugar. To obtain the sugar the cane has to be crushed, and although the juice so extracted can be consumed directly, it is normally processed further to produce crystallized sugar. Sugar is not an essential foodstuff. People can live healthy lives without ever consuming an ounce of sugar—probably healthier, since sugar promotes tooth decay and obesity. Yet in industrialized countries vast amounts of sugar are consumed in candy and desserts, and in a hidden form in almost every type of industrially prepared food.

Sugar (sucrose) is not a common product in the plant kingdom. Nevertheless from early times people have identified and cultivated the few plants that produce it. The sugarcane and the sugar beet provide more than 90 percent of the sugar we consume today. Since processed sugar is compact and desirable and keeps well, it is a valuable commodity. It has been traded from the time of the earliest available records. Its history exemplifies the transformation of agriculture from a way of obtaining food to a process of accumulating capital. Sugarcane cultivation is linked with another momentous social change produced by agriculture, the development of the plantation, a forerunner of the modern factory system.

Growing, extracting, and processing sugar are all labor- and capital-intensive operations. Sugarcane is a demanding crop whose cultivation is associated with soil degradation and destruction of forestland in most of the places it has been grown: India, Mediterranean islands such as Cyprus and Sicily, and in the Americas, the Caribbean, and Brazil. It is best grown in large estates that make efficient use of facilities. Throughout history it has been difficult to attract the labor force needed for the grueling work of cultivating, harvesting, and processing sugar. More often than not, people have been coerced to work in sugar plantations. The sugarcane is forever linked to slavery, exploitation, misery, and human suffering. This crop is directly responsible for the elimination of the native inhabitants of the Canary Islands, the *guanches*, and for the involuntary transportation of thousands of Africans to the Americas. It is ironic that the crop that sweetens our meals and takes away the bitterness in our drinks, the crop that we associate with pleasant sensation, is also the one that throughout history was the cause of extensive human misery.

Using the energy of sunlight green leaves synthesize glucose, the most basic sugar molecule, from carbon dioxide and water. This process, known as photosynthesis, is the most fundamental function of the plant and is what makes life on earth possible. The sweet taste of many fruits is due to glucose and its close relative fructose. Sucrose, the raw sugar we get from sugarcane, consists of one molecule of glucose and one of fructose. We shall use the term *sugar* to refer to sucrose. Synthetically, it is a simple step to go from glucose to sucrose, but few plants do. Instead most plants make starch, a large molecule formed by stringing together many glucose molecules. All the plant foods we eat contain starch. Roots and tubers like potatoes and carrots are especially rich in it, as are seeds, particularly cereal seeds. Fruits and vegetables such as tomatoes, eggplants, and apples also have starch.

Starch, unlike sugar, is an essential food element. It provides the bulk of the calories we need to function. It is the fuel that runs the human body. Our cells break down starch into glucose molecules, which then are taken by the bloodstream to all the tissues of the body. Since starch is a large molecule, the body breaks it down slowly, providing a steady supply of energy between meals. Sugar, on the other hand, is immediately broken down into glucose and absorbed into the bloodstream. It gives an energy shock and little else—akin to pouring ethyl alcohol into a cold engine in the winter to get it started. Starchy foods, moreover, are not just raw starch but contain small quantities of other nutrients such as proteins and vitamins, while refined white sugar is nothing but sucrose.

Plants store starch for times in their life cycle when they need to expend more energy than they produce—for example, when a plant grows new leaves after a cold winter, when a desert plant emerges after a prolonged drought, or when a new plant germinates. Starch has the advantage over sugar of packing more energy per unit of space. Nonetheless, plants store sugar for similar reasons. Sugarcane uses its stored sugar to form flowers and seeds.

As a raw food, sugarcane is not very important. The edible part, the juice, is present in the internodes of tall and semiwoody reeds. To get at it, one or more stalks have to be cut down, broken into pieces, and pressed. The sugary juice is bulky and hard to transport, and it does not keep well. These drawbacks can be avoided by boiling down the juice. Once most of the water evaporates and the syrup cools, the sugar in the juice crystallizes. This process takes time and consumes fuel, and is unlikely to appeal to farmers solely interested in growing food plants. However, the resulting product has none of the disadvantages of the raw juice and people are willing to pay good money for it. Sugar has many additional advantages as a commodity. Not only does it store and transport well, but because its preparation requires so much labor and equipment (capital) as well as expert knowledge, the lucky few who master the art can enjoy high profits.

Sugarcane is a perennial plant and in its natural state will produce several tall shoots from the base. After the shoots bloom they die out, and new ones growing from the base replace them. To raise sugarcane commercially, farmers place setts, or pieces of cane a meter long with three to five internodes, along a furrow. Then they cover them with soil. From each node a new shoot develops. Growers do not allow sprouts to bloom but harvest them just before flowering begins by cutting the shoots close to the base with a knife, machete, or machine. New shoots called ratoons will grow from the cut bases. This second crop does not produce as much sugar as the first, and each succeeding one yields still less. According to soil and climate, sugar growers repeat this process (which in theory can go on indefinitely) two to five times, until yields are too low to make it commercially viable. Then they plow the fields, fertilize them, and plant them again.

There are five species of sugarcane, all belonging to the genus *Saccharum*. We know three of these, *S. officinarum*, *S. barberi*, and *S. sinensis*, only from cultivation; the other two, *S. robustum* and *S. spontaneum*, also grow wild. The exact ancestry of cultivated sugarcane is not clear. It probably comes from somewhere in Indonesia, where wild sugarcane still grows. Domestication likely occurred in New Guinea, and from there cane cultivation spread to India and China.

The sugar industry evolved first in northern India. The earliest information about processing sugar comes from Sanskrit manuscripts, which show that by about 300 B.C. Indian farmers were already manufacturing sugar in India, at least through the first two stages of extraction and crystallization. By A.D. 300 Indians were manufacturing crystalline sugar, and sugar growing was important in the Ganges Valley.[1] It appears that Indians used sugar as medicine rather than food, suggesting that the cost of production was high.

The Chinese learned to cultivate sugarcane by the time of the late Chou and early Han dynasties, circa 200 B.C.; by the time of the Sung Dynasty (A.D. 960–1279) sugar growing was widespread in China.[2] The Chinese used sugar in the preservation of food and shaped it into images of people or animals for candy.

For a long time historians believed that Alexander the Great introduced sugarcane into Europe. From Greek and Roman sources it is now known the Europeans were aware of sugarcane from travelers' accounts only. Although some samples may have been brought back by travelers, Europeans did not grow sugar themselves. In the first century A.D. Dioscorides, a Greek physician in the Roman army, wrote a book on medicinal plants that became a standard for centuries. In it he describes sugar thus: "And there is a kind of concrete honey, called sugar, found in reeds in India and in Arabia the Happy, like in consistency to salt, and brittle to be broken between the teeth, as salt is. It is good for the belly and stomach being dissolved in water and so drank; helping the pained bladder and the veins."[3]

It was the Arabs who introduced sugarcane growing into the Mediterranean basin. The plant was cultivated in Persia by A.D. 600 and in Egypt by A.D. 700. To grow sugarcane in Egypt required irrigation. Although Egyptians had irrigated their fields since ancient times, irrigation was applied primarily to cereal crops growing in winter and spring. Perennial sugar had to be irrigated also in summer, and the periodic floods of the Nile were no longer sufficient. Techniques had to be developed to pump water from the Nile when it was low. The Arabs learned to do this, and it allowed them to grow not only sugarcane but other tropical perennial crops that they introduced to the world, such as bananas, citrus, and mangoes.[4]

The Arabs took their sugarcane and irrigation technologies with them when they conquered northern Africa, Spain, and Sicily during the seventh to ninth centuries. After the Normans reconquered Sicily and the First Crusaders the Levant, Europeans learned how to grow the new crop and slowly a market for sugar developed in Europe. Sugar was exorbitant, all through this period a luxury available only to the rich. As in India, it was considered not a food but a medicine, honey being the universal sweetener.

Sugarcane is a tropical crop. It is killed by frost and requires heat and a lot of moisture for optimal growth. These conditions are absent in the Mediterranean. Why did people grow it in the Mediterranean region for over seven hundred years?

The simple answer is market. Since there were people willing to pay a very high price, producers were willing to grow sugarcane even under marginal conditions. This was a departure from the primarily self-sufficient agriculture of the times. The second reason was that the high cost of transportation protected farmers in the Mediterranean market from competition from producers outside Europe. Sugar was grown in India, southern China, and Southeast Asia under much better conditions, with better yields and lower production costs, but communication between Europe and the Far East at that time was by land, and given the bulk and weight of sugar, transportation would have made its shipping prohibitively expensive.

Mediterranean cane was a hybrid between *S. officinarum* and *S. barberi*, originally introduced from India and later taken to South America. Not until recently were other species cultivated in the West. Because it did not grow in the cold winter months, Mediterranean farmers planted it in early February or March and harvested it the following January. To speed up growth of a newly planted crop and protect it from the cold, they often planted young cane in manured nursery beds and then transplanted it to the fields. Arab cultivators experimented to obtain the ideal planting density. Early on growers appreciated the importance of fertilization and spread cow, sheep, horse, and donkey manure generously on cane fields.

The method of extracting juice from cane consisted of two steps. The cane was cut into small pieces, then put in a mill similar to the one designed for grinding seed crops. The mill extracted only part of the juice. Residues were further pressed in a beam or screw press like the one used to press grapes. Sometimes, instead of the usual grindstone, a mill had a large, rotating stone wheel that turned on its edge around a central axis, to which it was connected by an axle. After extraction, the juice was boiled down in a large kettle or cauldron. Impurities that surfaced were skimmed off. When the juice reached the point of crystallization, it was poured into clay molds shaped like inverted cones, which produced traditional "sugar loaves."

Besides sucrose, the concentrated juice contained a noncrystallized portion known as molasses, mostly glucose. The molasses moved toward the bottom of the mold as it cooled and drained through a hole. As it seeped away it carried some of the remaining impurities that gave the raw sugar its brown color. When workers removed the sugar from the mold after cooling, it would have a color gradation, lighter at the upper base of the cone than at the lower tip. The whiter sugar, which commanded a higher price, was often cut off, heated, and recrystallized to increase its purity. There are accounts of up to four such boilings. Another form of purification involved the use of clay.

Given the climatic constraints of the Mediterranean and the technical limitations of the times, there was not much that sugar producers could do to improve yields or reduce costs by purely technical means. The industry instead focused on reducing the costs of labor. Slavery had survived in the Mediterranean since Roman times, tempered by Christian precepts. To it had been added new forms of compelled labor such as serfdom. The wars between Christians and Moslems were the main source of new slaves, but it was not until Portugal opened up the Atlantic route to Africa in the fifteenth century that slavery became central to the sugarcane industry. This new source of labor was responsible for its phenomenal growth.

As sugar estates grew in size and labor requirements increased, the production system known as the plantation took form. Plantations existed in Cyprus and Crete in the twelfth century A.D., but Sicily is where they became a model for the future Atlantic sugar industry. Typically an absentee proprietor owned a plantation, often in partnership with rich merchants who provided capital. Planting, milling, and refining operations were usually run separately, with an overseer in charge of each. Labor was mostly, if not exclusively, performed by slaves, with time entirely by African slaves. Planting, weeding, and especially harvesting were the most physically demanding and least prestigious operations. Slave gangs of men, sometimes chained, did that part of the work. Both men and women ground the cane, a dangerous operation in which arms often got caught in moving machinery. It was crucial that the cane be ground soon after cutting, since sugar content decreases quickly after cutting.

The most prestigious and important operation was boiling down the sugar. The sugar master supervised this process. Whether slave or free, he was well paid for his efforts. Assistants filled the kettles with juice, skimmed off impurities, and poured the sugar into molds. A stoker kept the fires lit under the cauldrons. The last step was refining and packaging the sugar.

Thus the sugar industry developed the hallmarks of commercial and capitalist agricultural production: concentration on one crop, production geared exclusively toward a market, specialization, and rationalization of production aimed at maximizing capital yields. In the process, it adopted exploitative labor practices and ecologically unsound methods, such as cutting down forests to obtain fuel.

These developments took place when most European agriculture was still traditional and geared to the production of cereals and foodstuffs. The sugar plantation in the Mediterranean was the opposite of the feudal estate. The estate grew plant species primarily to satisfy the food and clothing needs of its inhabitants, with a small surplus for nearby city markets. The sugar plantation concentrated on one crop for a specialized market, the courts and cities of Europe. In the estate labor relations were two-sided: Both lord and serf had obligations toward each other (although the relationship was not symmetrical). On the plantation the relation was decidedly one-sided: Slave laborers were seen as property and treated as an investment. The estate aimed at ensuring stability and security, while the plantation's purpose was to produce profits and accumulate capital. The lord traditionally lived on the estate; the sugar capitalist resided in the city.

With the development of the sugar industry, the character of agriculture in the Mediterranean basin became more commercial. In classical and early medieval times agriculture in the region centered on foodstuffs. Although land was privately owned and farmed for a profit, it produced for a local market. Rome did import wheat from Egypt, the Levant, and Carthage, but part of the imports were coerced tribute.

The oil and wine industries, developed to supply larger and more selected markets, were fairly profitable. But although wine today is not
an essential foodstuff, it was considered as such in classical times. In
any case, most oil and wine were consumed locally. Grape vines and olive trees and the paraphernalia required to extract and prepare wine
and oil represented a substantial investment, and these were the first
agricultural industries in the West.

Sugar represented a further intensification of agriculture as an industry. Unlike the wine and olive industries it required sizable capital
investment in irrigated land, equipment, and labor, and extracting
and refining sugar required special knowledge.

At first, sugarcane was grown in peasant-owned plots. These varied
in size. As time went by, land became concentrated in fewer and fewer
hands, and the system of plantations run with slave labor came into
being. In some parts of the Muslim world, for instance Egypt, sugar was
grown on state-owned land apportioned to army officers. Since sugar
growing required a great deal of capital, for the first time in history
there were outside capitalists who invested in an agricultural industry
without taking direct part in its management.

The center of gravity of the sugar industry moved from East to West
during the period of Arab influence in the Mediterranean. In the early
period, the Levant and Egypt were the major producers and exporters
of sugar to the rest of the Muslim world and Christian Europe. By 1300
the center of gravity had moved to Cyprus, Crete, and Morocco, and
by 1400 to Spain. War and strife in the East, and possibly the exhaustion and negligence of producing lands, were responsible for this shift.

Sugar cultivation in the Mediterranean basin had always been marginal because of cold winters and dry summers. During the five hundred
years after the Arab invasion, a conjunction of factors made it possible
to grow sugarcane there. These were the intensification and perfection
of summer irrigation, the existence of a market willing to pay a high
premium for sugar, the relative isolation of Europe at the time, and finally a period of warmer weather that reduced frost damage. But by the
fifteenth century changes occurred that signaled the end of sugarcane

as a Mediterranean crop. The region entered a period known as the Little Ice Age that lasted well into the nineteenth century. Mostly because of internecine war, Arab irrigation works were not maintained, resulting in lower yields and higher production costs.[5] Advances in seafaring vessels made transportation cheaper. Finally, sugar from the Atlantic islands appeared in Europe. Production there was more reliable and cheaper than in Europe.

Meanwhile Christian Crusaders were keeping up pressure on the Arabs in Spain. In 1212, Christian knights in the service of the king of Spain defeated Arab forces at the Battle of Navas de Tolosa. Over the next century they occupied all the fertile Guadalquivir Valley, where much sugar was grown. In 1492 the rulers of Spain, Ferdinand and Isabella, conquered the last Arab kingdom in Spain. With that event, sugar growing moved definitively to the Atlantic, specifically to the recently discovered Azores, the Canary Islands, Madeira, and São Tomé.

Portugal, at the urging of Prince Henry, "the Navigator," initiated at the beginning of the fifteenth century a program of explorations. The first discoveries made by the Portuguese were the uninhabited Azores and Madeira. Some time later Spanish navigators discovered the Canaries, which lie south of the Iberian Peninsula. These Atlantic islands have a warmer and milder climate than the Mediterranean basin, and on the windward side they are extremely wet. With their discovery began a new chapter in the history of sugar.

At the urging of Prince Henry, Portuguese settlers colonized the Azores and Madeira around 1425. They brought their skills as cereal growers, and soon the Azores were exporting wheat to the mainland. Genoese merchants saw the possibility of planting sugar in the islands. Two systems of agriculture then started competing for land and resources: cereal agriculture in small plots cultivated by freeholders, and sugar plantations run mostly with slave labor. In the colder and somewhat drier Azores, cereal agriculture won over sugar. In Madeira (and in the Canary Islands and São Tomé) the story was different. Here farmers could grow sugar year-round without irrigation, producing yields higher than anywhere on the continent. So sugarcane took over all the suitable land, and by the middle of the century Madeira was im-

Map of the western Mediterranean and the Atlantic, showing the principal sugar-growing sites mentioned in the text.

porting wheat. Madeira sugar began arriving in Portugal about 1450; by the end of the fifteenth century sugar from the Atlantic islands undersold any produced on the continent.

However, the life of the Atlantic islands sugar economy was brief. Already on his second trip to the New World, Columbus took sugarcane to grow in Hispaniola and Cuba. Although sugar was to prosper there and later also in Mexico and Peru, it would remain mostly a crop for local consumption in colonial Spanish America.

Brazil was a different story. Disappointed at not finding precious metals, the Portuguese settlers there turned to growing sugar. All along the coast, from the mouth of the Amazon to São Paulo, sugar plantations sprang up. It was in the Reconcavo area of Bahia, in northern Brazil, where the sugar plant found ideal growing conditions. Sugar was to dominate the economy of the colony for more than three hundred years until eclipsed by another invader from overseas, coffee. Sugar would give colonial Brazil its character, decide the composition of its population, and dominate its politics. Bahian sugar was eventually produced in such quantities and at such low prices that it transformed what had been a luxury and a medicine into what it is today, a foodstuff craved by most humans in one form or another. From the seventeenth century on, it would be subject to strong competition from the Caribbean sugar islands, which eventually dominated the industry.

Pedro Alvares Cabral discovered Brazil in 1500, supposedly by accident while en route to India, though modern historians question this account.[6] A colony was settled. The early years were rough. There were no finds of precious metals. The native Indians had little to offer and after first accommodating the newcomers became quite hostile. French sailors and pirates competed for the trade in Brazil wood, which was used as a dye. Although there is evidence of sugar production in Brazil already in 1519, the industry was then very small.[7]

In 1533 the Portuguese crown allotted different areas of the Brazilian coast to "donatary" lords. Under this arrangement, the crown bestowed on the captain general of an area broad powers of jurisdiction, taxation, and administration, which were reserved in Portugal for the highest level of the nobility. The objective was to promote economic development. The captain general was to recruit colonists and capital to develop the new colony and to keep out intruders, especially French and Dutch.

All the newly established captaincies, from São Vicente (present-day São Paulo) in the south to Pernambuco in the north, planted sugar. The cane was introduced from the Portuguese Atlantic islands or from a neighboring colony. Settlers built small sugar mills in several places. The Portuguese coerced Indians into working in the fields and the mills, following the Mediterranean pattern. This resulted in several uprisings—Indians burned fields and sugar mills at several sites, including Espírito Santo, north of Rio de Janeiro—and most of these first efforts were temporarily abandoned.

Two captaincies escaped failure. They were São Vicente and Pernambuco, situated at the extreme southern and northern ends of the area of Brazil then claimed by Portugal. In São Vicente good relations with the Indians and sufficient capital determined the success of the enterprise. However, sugar in São Vicente was primarily a local crop; only occasionally was some exported to Europe. The major problem was distance, São Vicente being twice as far from Europe as Pernambuco.

Pernambuco proved to be the most successful of the donatary captaincies established in Brazil, largely owing to the donatary Duarte Coelho. He moved to Brazil with his family, determined to make a success of the venture. They established good relations with the Indians. Several colonists married Indian women, including Jeronimo de Albuquerque, the brother-in-law of Coelho. Sugarcane was planted in 1542. By 1550 there were five sugar mills and by 1580 sixty-six. By the end of the century Brazilians were exporting sugar to Europe, mostly in Dutch ships. Labor was a problem from the beginning, and already in 1542 Coelho petitioned the Portuguese crown for the right to import African slaves.

Bahia, to the south of Pernambuco, had a false start. The donatary, Francisco Pereira Coutinho, proved to be a weak leader and poor administrator. Soon the colony got into trouble with local Indians, and the fighting that followed led to Pereira Coutinho's death. But the Bahia de Todos os Santos and the area surrounding it, known as the Reconcavo, was ideal for growing sugar: The temperature and soil were perfect and there was plenty of rain. Thus in 1548 the Portuguese king, Dom Joâo III, decided to establish a crown colony in Bahia and sent a large expedition under the direction of Governor Tome de Sousa, who proved an able administrator. He founded the city of Salvador da Bahia do Todos os Santos, which became the capital of Brazil until 1763. Land grants were given to planters willing to grow sugar and build mills. Soon a dynamic sugar industry developed in Bahia.

Large landholders had *engenhos*, large sugar mills, which were the center of the industry and of Bahian society. An *engenho* was associated with a large plantation, but the mill also ground cane for smaller sugar growers. Slave labor, almost entirely African, ran the mill. Native Indians could not adapt to the hard and routine work of the plantation and either ran away or revolted. They were also susceptible to European diseases.

The tropical climate of Bahia allowed year-round cultivation and extraction. Mills operated usually around the clock ten months out of the year, from early September to the end of June. It took about eighteen months for cane, planted in the traditional way, to be ready for harvest. After the first crop, the new growth or ratoon took only twelve months. Planters grew three or four such crops, each successive ratoon crop yielding less than its predecessor, before replanting the field.

The first operation was planting by gangs of slaves. The field was then weeded with machetes or hoes. When the cane was ready, harvesters cut it by machete or knife close to the base, stripped the leaves, and tied the stalks into bundles. The bundles of cane were transported by oxcart (or by boat if the mill was far away) to the *engenho*.

Milling techniques had advanced over those the Arabs introduced into the Mediterranean. The grinding stone that required the cane to be chopped was replaced by the horizontal two-roller mill, in turn replaced by the vertical three-roller mill. The two-roller mill consisted of two rotating stone cylinders placed sideways, one on top of the other. The uncut cane was jammed and crushed between the two rollers and the juice collected at the base. Turning the heavy rollers was hard work done by hand or with oxen or horses. The new mill removed the juice more efficiently than the grinding stone designed for grain, and the process eliminated the step of cutting cane into small pieces.

The next progression was the application of water power to move rollers, permitting the building of larger mills that could grind more cane. The three-roller mill was probably invented in Brazil.[8] Three rollers were placed vertically and side by side. Animal or water power turned the middle one, while the two lateral ones turned in opposite directions as a result of friction with the first. Operators stood on either side of the mill and passed the cane back and forth until all the juice was squeezed out. The cylinders were made of hardwood rather than stone.

Sketch of a three-roller mill pulled by a donkey, still in use in the Venezuelan Andes.

The favorite source of power was hydraulic. Slaves, usually women, spent ten to twelve hours a day putting tons of cane through these mills. It was dangerous work. If a finger got caught in the rollers, there was no way of stopping the mill fast enough to avoid crushing the whole arm. Accidents were often fatal.

The juice was then sent on for evaporation. Advances had also taken place in this part of the manufacturing process. Wooden pipes carried the juice to a large vat. Workers ladled or poured it into the cauldrons. The cauldrons were made out of copper, and there were elaborate theories about the best shape for distributing the heat of the fire. In place of the old single cauldron, *engenhos* used a series of three to seven kettles of diminishing sizes placed side by side and heated by a common fire. This was called a train. Under the direction of the sugar master, workers ladled juice from one kettle to the next to keep them all at the same level. The first kettle in the train was the largest and contained the watery sugar. As water evaporated from the mixture, the increasingly concentrated juice was transferred to the smallest vessels, which were closer to the fire and therefore hotter. The first kettle was refilled with freshly squeezed juice. When the master determined that the sugar was ready, workers poured it into clay molds and allowed it to cool. The by-product, molasses, was either consumed locally or used for making rum.

To refine the sugar, workers poured a mixture of clay and water on the tops of the molds and allowed the clay to drain through. The clay trapped any remaining molasses and other impurities, and left behind an almost pure white sugar, especially at the top. The last step was to remove the sugarloaf from the mold and divide it into layers of different-quality sugar, depending on its color. The first-grade sugar was packed in large wooden crates weighing between 50 and 250 kilograms (they kept getting larger with time) and shipped to Portugal.

The seventeenth century represented the golden era of Brazilian sugar. Portugal had a virtual monopoly on the European sugar trade,

A section of a "Jamaica train." Note the decreasing size of the cauldrons and the fire below the last cauldron in the series. Sugar juice is poured at the left and as it warms and the water evaporates, the increasingly concentrated juice is transferred slowly to the right until the sugar is ready to be poured into molds, where it will crystallize as it cools.

and although not problem-free the industry grew steadily. Sugar was shipped from Bahia to Lisbon, and from there it was shipped to northern Europe, usually Amsterdam. The Dutch refined much of the sugar again and used some of it in the manufacture of chocolate, the cocoa beans for which arrived from America, especially Mexico and Venezuela, through Spain.

The prevailing economic theory of the times was mercantilism. All countries tried to sell to others while attempting to reduce as much as possible their own consumption of imported materials. In 1580 the Portuguese and Spanish crowns were united under Philip II. The Spaniards, engaged in a struggle with their rebellious provinces in the Low Countries, interrupted the flow of sugar to Holland. The Dutch, heavily dependent on this trade, retaliated by occupying Pernambuco in 1630. They also succeeded in occupying Bahia but were driven out after one year. The Dutch stayed in Pernambuco for twenty-four years before the Portuguese settlers forced them out. Then the Dutch took their newly acquired knowledge of sugar cultivation north, where they set up plantations in Suriname.

The English and the French, who had wrested possession of some of the Caribbean islands from the Spaniards, also started to grow sugar, especially in Barbados, Jamaica, and Haiti. With more capital and a bigger market, they soon became strong competitors of the Brazilians and eventually overtook them. The eighteenth and nineteenth centuries saw the pinnacle of the sugar industry in the Caribbean, based on efficient but ruthless exploitation of slaves. The Caribbean islands depended on imports of foodstuffs from the English colonies in North America, slaves from Africa, and manufactures from England. They produced only one commodity, sugar, which Europe craved.

The English traders in the Caribbean developed what historians call the triangular trade. They shipped sugar produced in Barbados or Jamaica to England. With the proceeds of the sale of sugar they bought salt, firearms, cast-iron bars, cloth, gunpowder, shot, and alcohol and shipped these commodities to the west coast of Africa. These products were bartered for slaves. Native chiefs from coastal tribes usually obtained the slaves by raiding tribes from the interior of Africa. The slaves were transported under abominable conditions, often chained, to the Caribbean and sold. Then the process started all over again, ships being loaded with sugar for the passage to England. Profits were handsome in all these transactions. However, losses were also great, from shipwreck, pirates, mutinies, and disease, especially in the leg from Africa to the Caribbean. As time went by the seaworthiness of ships improved, and as the Royal Navy established its command over the Atlantic losses decreased considerably.

By the end of the eighteenth century, England controlled over half of the sugar trade between America and Europe, and the Portuguese industry in Brazil was in clear decline. England consumed more than 50 percent of the sugar it imported, while the remainder was reexported to the continent. As much as 25 percent of all of Britain's merchant marine may have been involved in the triangular trade, not to mention some quarter of a million workers in England and the thousands of slaves on Caribbean sugar plantations.[9]

The world of the nineteenth century was different from what it had been before the Napoleonic wars. Industrialization brought with it steam and the ability to make large machinery cheaply and efficiently. Steam also changed transportation, bringing the cost of shipping down considerably. Belatedly, the world's conscience turned against slavery. Simultaneously the populations of Europe and North America began burgeoning, and their living standards improved. The sugar beet was domesticated in Europe, introducing a second high-yield sugar plant. All these circumstances had a strong influence on the Caribbean sugarcane industry. Gone was its monopoly, and gone (or soon gone) was its labor supply. However, the industrial revolution brought with it new ways of solving many of these problems.

Small-gauge mini-railroads sped the movement of cane from the field to the factory. Trucks later replaced them. Better knowledge of fertilization, and the use of improved varieties instead of the old creole hybrid stock brought by Columbus, increased yields. The most drastic change was in extraction and processing. Automatically fed steel roller mills, which squeezed out every ounce of extractable sugar in one pass, replaced the three-roller hand-fed mill. And sugar juice was no longer boiled over an open fire; this step now occurred under vacuum in precisely controlled, steam-heated boilers. Charcoal filters were adopted to purify the product. [10]

Sugarcane was not the first industrial plant in Western agriculture; olives and grapes preceded it. However, owing to the difficulties of growing it in the Mediterranean environment as well as extracting and processing it, sugarcane required much more intensive capital and human investment than olives or grape vines. The European elite of the tenth century and later had a penchant for sugar and were willing to pay for it. Thus was the ground laid for a revolution in farming. For the cultivation of sugarcane remade agriculture. No other plant was as influential in the transformation of farming for food into farming for trade and capital accumulation. No other plant was the cause of so much human suffering. And few plants have been as ecologically destabilizing. Sugarcane drains nutrients from the soil in which it grows. More important, the vast amount of fuel required to process sugar leads to a great deal of deforestation and accompanying soil erosion wherever it is grown, be it the Mediterranean, the Atlantic islands, Brazil, or the Caribbean.

Exchanges

COLUMBUS WENT TO sea in search of the "Spice Islands," the source of clove, black pepper, and cinnamon that had become such important commodities for Europeans in the fifteenth century. Although at the time most Europeans were peasants and probably had not heard of the Spice Islands, much less tasted their reputed products, the European elite had established a "world" market for high-value agricultural commodities such as sugar and spices. Amsterdam was the center of that market.[1] Columbus never got to the Spice Islands, but the land he discovered, and the crops that the native people in the Americas had domesticated, would be far more important to the history of Europe and the world than the spices he sought. The early

Spanish explorers, including Columbus himself, brought back from the New World all manner of exotic plants, animals, and artifacts. At a time when travel, even over short distances, was rare because of its danger and cost, these curiosities held the fascination that in modern times moon rocks have inspired. Edible plants had the added attraction of their potential economic value. For a century after Columbus's voyage, European explorers, clerics, and scientists would carry plants and animals to and fro across the Atlantic in a transfer that has been called the Columbian exchange. [2]

Among the plants that America contributed to the world are avocados, bananas, the common bean, manioc, cacao, maize, upland cotton, papayas, peanuts, peppers, potatoes, squashes, tobacco, and tomatoes. In turn, Europe and Africa contributed their plants (most originally from Asia) to America, including coffee, grapes, rice, soybeans, sugarcane, and wheat. Two American crops and one African crop figured prominently in the exchange—potatoes, tobacco, and coffee—and had a profound impact on the social structure of the West. They were major contributors to the life-style of urban society.

The most important plant components in the human diet, after seeds, are roots and other underground tissues. We call these root crops, even though only some are in fact roots (carrots). Most are bulbs (onions), tubers (potatoes), or rhizomes (taro). As is true with seed crops, a few species account for the bulk of the underground plant tissues people eat. They are potatoes in the temperate zone, manioc (also known as yuca, cassava, or mandioca) in the South American and African tropics, and sweet potatoes, yams, and taro in Asia and Oceania. Other important root or rootlike tissues are beets, carrots, and members of the onion family (onions, garlic, shallots, leeks). Potatoes, sweet potatoes, some species of yams (there are also species of African and Asian origin), and manioc are all American plants.

Underground storage organs such as the potato function like seeds. Plants that produce bulbs, tubers, or rhizomes usually live in harsh environments, with either a dry or a cold season when plant growth is not possible. The aerial parts of the plant die, and it survives the unfavorable season in the form of underground organs that reserve nutrients, mostly starch, fill. When the unfavorable season is over, a new plant grows out of the subterranean structure. Often not just one plant but many appear.

Being full of nutrients, the underground tissues are good food for animals, including humans. To protect themselves from predators, most wild tubers have some kind of poisonous or bad-tasting (usually bitter) substance, which has to be removed before eating. Plant breeders have eliminated these substances from most domesticated species. People grow other plants such as garlic because of the special taste these defensive substances impart to the bulb.

Root crops have advantages and disadvantages over seed crops. The major advantage is productivity. In general a root crop produces five to ten times more bulk than a seed crop. A field of potatoes feeds more people than an equivalent field of wheat. On the minus side, root crops have less protein and more starch than seeds, although not as little protein as some people imagine. With more water per unit weight, they don't keep as well as seeds. Being bulkier, they are also more expensive to transport. Therefore root crops tend to be local ones eaten by the poor. People in the lowland tropics don't eat many potatoes. We in the temperate zone eat only a little manioc, mostly as tapioca.

America gave the world some of its most important food plants. Excepting maize, probably none is as important to the modern diet as the potato. The history of central and northern Europe in the last two centuries and that of the industrial revolution are closely tied to this plant. So one would expect there to be extensive knowledge of its introduction and expansion. That is not the case. There is a great deal of confusion, and fiction is as common as fact in many existing popular accounts regarding the diffusion of this crop.

The potato belongs to the Solanaceae, a large plant family. Other cultivated members of this family are tomatoes, eggplants, and red, green, and chili peppers. The potato genus, *Solanum*, is one of the largest genera in the plant kingdom, and its species are very diverse. The formation of tubers characterizes one group of species. Wild potatoes are smaller than cultivated potatoes, one to three centimeters in diameter. Tuber-forming species of *Solanum* are native to South America and grow wild from Chile to Colombia. They are for the most part diploid, and many of them can be crossed.

The cultivated potato is not a diploid but a tetraploid plant. It does not occur in the wild. All evidence points to the highlands of Peru and Bolivia as its place of domestication. The native farmers of these areas cultivate an astonishing number of varieties and species of potatoes, some of which are diploid rather than tetraploid. One hypothesis is that in prehistoric fields, a diploid ancestor of the potato crossed with another weedy species (or perhaps two different species were being cultivated together) and the hybrid then doubled its chromosomes. Such a process occasionally happens spontaneously in this genus. According to one theory, our potato originated from a cross between the cultivated diploid species *Solanum stenotomum* and the weedy *S. sparsipilum*.[3]

Another characteristic of Solanaceae species is that they produce alkaloids and therefore are usually poisonous. The potato's green tissues (leaves, stems, and green tubers) are poisonous. The tubers of many wild species are bitter. Early farmers selected plants with sweet tubers.

If the potato is indeed native to the highlands of Bolivia and Peru, it was probably first cultivated before 2000 BP, possibly as early as 8000 BP. It and maize became the staples of the Incas. It was the first root crop to become so basic to a civilization. The Incas had their capital in Cuzco, west of Lake Titicaca, in a high mountain valley. The valley was one of the more fertile and humid ones in the area, but still dry by world standards. Among the most remarkable aspects of the Inca civilization was its system of crop cultivation. It was based on two major food crops, maize and potatoes, and two fiber sources, cotton and llama wool. Maize and cotton are primarily tropical crops that require ample water for growth, while the potato is a crop of cooler climates that can survive with less water. The differing requirements of these crops enabled the Incas to plant every cultivable area in this inhospitable region with the help of irrigation.[4]

The Incas developed an elaborate system of irrigation canals in pre-Columbian South America, bringing water from tens and even hundreds of miles away. They also developed an intricate system of cultivation. They grew a mixture of crop species and varieties in the high Puna to ensure that some crops would yield even in the coldest and driest years. The Incas also perfected a storage system for agricultural surplus. Their elaborate granaries stored many products, potatoes and maize being the two most important.

One disadvantage of tubers is their tendency, because of a high moisture content, to rot when stored for a long time. The Incas learned to treat their potatoes to prevent this, a process that today we would call freeze-drying. To make *chuño*, peasants exposed potatoes (usually of the bitter varieties) to the freezing night temperatures of the high Puna. Frost killed the cells in the tubers. As a result the cell membranes became permeable and permitted the movement of water. During the day water was squeezed out of them, usually by trampling underfoot. In the dry atmosphere of the Puna the water evaporated. This process continued for several days and nights until the tubers had shrunk to dry, small structures that would keep for years. The inhabitants of the Puna still make *chuño* today.

Already in Inca days potatoes were the food of the poor. The Incas paid less attention to their cultivation, growing them in the more marginal land of the lower Puna without fertilization and without irrigation. Since the Spaniards appreciated neither the poor nor this new and unknown crop, they left a meager record about Inca potato cultivation compared with what they wrote about maize. However, potatoes were tremendously important to the Inca economy. One index of their significance is that in the upper reaches of the mountains, people measured time in pre-Columbian days by the potato's boiling time.[5]

The adoption of American plants by Europeans was slow. In the sixteenth and seventeenth centuries, most European farmers were still producing primarily for themselves, with only small surpluses for the market. Crop failures could mean famine and starvation, and they were not infrequent. Under such circumstances there was not much of an incentive to experiment with unknown crops. Communication being poor, news of even successful crops took a long time to spread. And then farmers had to be equipped with more than just seeds or tubers. They had to have all the specialized knowledge required to raise a

The well-known drawing by Guzman Poma de Ayala (sixteenth century) of planting and harvesting of potatoes by Quechuas in Inca times. Note the use of the digging stick.

successful crop. Add to all that the general human tendency to be conservative in eating habits and it becomes clear why diffusion was slow.

The new plants from America did, however, attract the attention of botanists. They cultivated the plants in their gardens and published illustrated books about them known as herbals in which they described their real and fanciful attributes. Serious herbalists drew from nature for their specimens and tried to describe accurately the food and medicinal value of the species they studied. Less careful scholars copied from others, and either out of indifference or duplicity recorded some erroneous facts. One such relates to the potato.

The first persons to record the presence of the potato in Europe were Gaspar Bauhin of Geneva and John Gerard of London, both in 1596. Bauhin is an important forefather of modern scientific botany. In the *Phytopinax* he described the potato and gave it the scientific name still used today, *Solanum tuberosum esculentum*. (The last term was later omitted to fit the rule that plants must have only two-word names.) Gerard merely recorded the presence of the potato in a catalogue of the plants he had in his garden. A year later, though, he published the now-famous *Herball*, in which he stated that the potato had come from Virginia. This error became accepted and led to the allegation that Sir Walter Raleigh had introduced the potato into England or Ireland and hence Europe.

The most important early scientific account of the potato is by Jules Charles de l'Ecluse, known as Clusius. Clusius was an extraordinary scholar, cautious, accurate, and observant. In 1601 Clusius, who was then a professor at Leyden, published the *Rariorum plantarum historia*, in which he gave a detailed description of the potato based on plants he had grown from tubers sent to him in 1588. The potato, he said, was commonly cultivated in Italy, where it was "cooked with mutton in the same manner as they do with turnips and the roots of carrots." Italians used it to feed pigs. He wondered that the potato was not well known in other parts of Europe, but "now it has become sufficiently common in many gardens in Germany since it is so fecund." He equated the plant with South American potatoes from the vicinity of Quito in Ecuador, described in the *Chronicle* of Pedro Cieza.

Clusius's account establishes two interesting facts. At the end of the sixteenth century, a hundred years after the discovery of America, the potato was cultivated and consumed at least locally in Italy and was just being introduced into northern Europe. How did the potato get to Italy? Probably through Spain, which then controlled Italy from Naples south. Seville was the logical place for the introduction of the potato into Europe. All ships going between America and Spain in the sixteenth century had to use the port at Seville. The records for 1576 from the Hospital de la Sangre in Seville show that that institution bought potatoes as part of its normal provisions.[6] Supposedly potatoes were available in the local market in 1573.[7] Since it would take four or five years at least to multiply a few tubers into a salable crop, it is likely that Spaniards began growing the potato at the latest by 1570.

In Ireland, the potato was an important crop by early 1700, but it did not become central to the general European diet until the middle of the eighteenth or early nineteenth century. Why did it take some two hundred years for this crop to become accepted? Although no one knows the exact cause, two factors, one agronomic and one social, are probably responsible.

The potato is a late-maturing crop. Farmers plant it in the spring and the plants do not ripen until August at the earliest. Late varieties are not ready for harvest until after the first frosts have killed the plant. The length of the day (or actually the length of the night), which the plant senses chemically, determines when tubers start to form. Late-ripening varieties are "short day" because they don't start filling the tubers until daylight is less than fourteen hours. Early-ripening varieties begin filling their tubers when the days are longer. The original potatoes that came from the tropics were short day since days near the equator are never much more than twelve hours long. All early accounts of potato growing mention the lateness of their ripening.

It was not until breeders started experimenting with potatoes in the eighteenth century that early varieties were obtained.[8] The time when potato tubers start filling affects yield. If the tubers start filling late in the season, there will not be enough time to produce many before the plants die from frost. The farther north the more serious the problem, because frosts come that much earlier. A variety that can mature well in Italy does not necessarily do well in northern Germany or Scandinavia. Without long-day varieties the potato would never have migrated north.[9] Eventually mutations that changed short-day plants into long day took place and were selected by farmers.

Although the population of Europe was increasing in the sixteenth and seventeenth centuries, there was enough land at the time to feed everybody. This changed in the eighteenth and especially the nineteenth century. Farm units became smaller and the number of households assigned to them larger. The potato, with its greater yields, could help alleviate the looming food problem created by the shortage of land. It is understandable that the potato became first the food of the poor peasant, who would have felt most keenly the shortage of land. The conjunction of earlier-ripening varieties and population pressures in the eighteenth century probably accounts for the spread of the crop then.

The tobacco plant is a tetraploid hybrid not known in the wild. Its ancestors are probably two South American species, *Nicotiana sylvestris* and *N. otophora*, whose ranges overlap in an area of southern Bolivia and Peru where botanists believe the cross took place spontaneously. At maturity tobacco plants are between one and two meters tall. Normally farmers remove the flower buds when the plant starts blooming so that all the energy of the plant will go into leaf production. A single leaf can have a surface of more than 200 square centimeters, and a single plant can have more than 2 square meters of leaf surface.

At maturity the leaf contains 85 to 90 percent water. The first step in processing is to dry it. The next step is curing, which consists of different procedures according to the type of tobacco but may involve wetting and heat drying, or drying in a moist atmosphere. The leaves are then graded and sent to a factory for further manufacturing, at which time some additives (sweeteners, aromatic compounds, oils) may be added.

China is the principal producer of tobacco today, followed by the United States, India, Pakistan, Brazil, the Soviet Union, the Mediterranean area, and Bulgaria. But it was domesticated in the New World. When Columbus first landed on the island of Santo Domingo, he discovered that the inhabitants were inhaling the smoke of lighted rolls of leaves. Friendly natives offered him and his party dried leaves as a special gift. Gonzalo Fernandez de Oviedo y Valdez, who wrote one of the first histories of the discovery of America, noted:

The Indians have one [practice] that is especially harmful, the inhaling of a certain kind of smoke which they call tobacco, in order to produce a state of stupor. . . . The caciques employ a tube, shaped like a Y, inserting the forked extremities in their nostrils and the tube itself in the lighted weed; in this way they would inhale smoke until they became unconscious and lay sprawling on the earth like men in a drunken slumber. Those who could not procure the right sort of wood took their smoke through a hollow reed; it is this that the Indians call tobacco, and not the weed or its effects, as some have supposed. They prize this weed very highly, and plant it in their orchards or on their farms for the purpose mentioned above. [10]

Soon after, the Spanish found natives smoking cigars in Cuba and, on the mainland, artistically decorated pipes. The natives also used tobacco as snuff, applied it to wounds, and gave it to sick people to induce vomiting.

Smoking was an old custom among the Aztecs, as Aztec picture writings illustrate. The plant that the Aztecs smoked was *N. tabacum*, a member of the Solanaceae or nightshade family, to which potatoes belong. Members of this family produce alkaloids, some of which are deadly poisons and others systemic poisons, such as nicotine. In parts of North and South America and the West Indies, natives consumed another species, *N. rustica*. This was apparently the tobacco grown by tribes east of the Mississippi. In western North America, natives cultivated other species, such as *N. bigelovii*, *N. attenuata*, and *N. trigonophylla*. In Australia, the only place other than the Americas where species of *Nicotiana* are native, the aborigines used a few, among them *N. excelsior*, *N. gossei*, *N. ingulba*, and *N. benthamiana*. [11] They gathered these species from the wild.

The story of the introduction of smoking tobacco into Europe is full of legends about opposition and eventual acceptance. The Spaniards in the West Indies soon acquired the habit of smoking, as did the Portuguese, who came across the plant and the habit in Brazil. It caught on rapidly with sailors. The plant was taken to Spain and Portugal, but at the beginning the habit was confined to the ports trading with America.

The first English colony in Virginia failed, but the settlers stayed long enough to acquire the habit of smoking. They brought it back to England, and their mentor, Sir Walter Raleigh, became an inveterate smoker. Raleigh saw the commercial potential of tobacco. Apparently he himself introduced N. *tabacum* into Virginia from the Caribbean islands. It became an important crop in subsequent settlements.

Smoking turned fashionable in England during the reign of Queen Elizabeth. Tobacco imports from Virginia increased, and tobacco growing became the cash crop that permitted the survival of early English colonies in the southern Atlantic coast of North America. In 1612 John Rolfe, the husband of Pocahontas, became the first settler to export tobacco to England. So successful was the export trade that by 1620, 20,000 pounds of tobacco were exported from Jamestown to England.

King James I opposed smoking and tried to stop its use in England with the help of a pamphlet: "There cannot be a more base, and yet hurtful, corruption in a Countrey, then is the vile use (or rather abuse) of taking Tobacco in this Kingdome, which had moved me, shortly to discover the abuses thereof in this following little pamphlet. . . . A custome lothsome to the eye, hatefull to the nose, harmefull to the braine, dangerous to the Lungs, and in the black stinking fume thereof, neerest resembling the horrible Stigian smoke of the pit that is bottomlesse." King James's diatribe sounds modern. However, he could only make an emotional appeal, since the critical medical information was unavailable at the time. He followed his pamphlet with a stiff tax on tobacco, hoping to check its spread while enriching his coffers. This practice, like tobacco, spread all over the world, and today tobacco is one of the most heavily taxed agricultural products anywhere.

Meanwhile another kind of reputation for tobacco was spreading. On the basis of the Native American custom of treating wounds with tobacco juice, Europeans began to ascribe to tobacco a curative power against all kinds of disease. In a treatise on the healing power of American plants, Nicolo Monardes, a famous physician at the University of Seville, stated that tobacco cured coughs, asthma, headaches, stomach cramps, gout, and women's diseases.[12] Nobody was to promote the powers of the tobacco plant more than the French ambassador to the Portuguese court, Jean Nicot of Nîmes. He was only twenty-nine years old when, in 1559, he went to Lisbon to negotiate a marriage between fifteen-year-old King Sebastian of Portugal and sixteen-year-old Marguerite de Valois, daughter of King Henry II of France. Nicot befriended the scholar and botanist Damião de Goes, who introduced him to the tobacco plants he had growing in his garden. From these Nicot obtained seeds and grew plants in the garden of the French embassy.

Having heard of the ability of the plant to cure sores and tumors, Nicot applied the leaves to the wounds of some friends. After convincing himself of the efficacy of tobacco, he wrote home to well-placed friends in court exalting its medicinal properties. On his return to France he continued his efforts. His enthusiasm for tobacco led the English botanist John Ray in 1703 to name the genus of the tobacco plant *Nicotiana*. His name is also immortalized in the name of the alkaloid nicotine, which gives tobacco many of its properties, including the ability to build addiction in users.

It was not as a medicinal or ornamental plant that tobacco conquered the world. It was the custom of smoking, established first in England and the Iberian peninsula, that spread the plant. From England smoking spread to the Netherlands about 1580. Soon thereafter a Franciscan monk from Aix-la-Chapelle wrote to his superior in Cologne, "There are many Spaniards here who have brought their bad habits with them; in particular they have a new sort of debauchery which they call smoking. . . . Their soldiers do swagger about puffing fire and smoke from their mouths, and the silly people look on and gape with astonishment."[13] By 1620 smoking had become established in the Rhineland and in central Europe. The Thirty Years War spread the habit, as war has done ever since. Eventually it spread into Poland, eastern Europe, Turkey, and the Levant.

With the spread of tobacco came opposition to smoking by the clergy, who saw it as a wicked and godless habit. One of the first documents in this respect is the manifesto issued by the Council of Mexico on October 27, 1589: "But of regard to the reverence due to the Holy

Eucharist, it is hereby commanded that no tobacco in any form whatever be taken by clerics before reciting Holy Mass, or by any person before receiving Holy Communion."

If it was godless, it was also dangerous. To this day smoking accounts for a large number of fires caused by smokers who are careless or fall asleep while smoking. Authorities of all times have rightly been concerned about it. A number of German principalities and cities, including the kingdom of Saxony, banned smoking in the seventeenth century. Similar edicts followed in France, Switzerland, Sweden, Russia, and even China and Japan, some of them calling for the death penalty. However, as with similar prohibitions involving other habit-forming drugs, they had little effect, and tobacco smoking continued unabated.

Smoking was originally done with clay pipes. Because of the fire hazard and the smell of smoke, another custom sprang up, snuff taking. It became fashionable in the French court toward the end of the sixteenth century and from there spread to all of Europe, particularly in aristocratic circles. Cigar smoking was introduced at the beginning of the nineteenth century. Cigars were easier to smoke than pipes and soon became popular all over Europe. Everywhere decrees appeared banning cigars, but they were not enforceable. Soon thereafter cigarette smoking, a custom like cigar smoking introduced from South America, spread through Europe. Cigarettes quickly displaced other forms of tobacco consumption.

The word *coffee* is derived from the Arabic *quahweh*, originally a poetic term for wine. Since Muslims cannot drink wine, the term was apparently transferred to coffee. In Ethiopia, where coffee is native, it is called *bun* and the drink *bunchung*.[14] Coffee is also known as *mocha*, the name of the port in the Red Sea whence it was shipped to Europe in the sixteenth and seventeenth centuries. Like grapes, coffee was probably first used as a food and not as a stimulating beverage. African people used stone mortars to crush the ripe fruits and seeds. Then they mixed them with animal fat and fashioned them into round balls to be consumed on the way to the hunt or to war. The mixture of fat and caffeine gave them both energy and endurance. Today students staying up all night to study for an exam get similar results from consuming coffee and hamburgers.

Africans used coffee berries to prepare a fermented alcoholic beverage and also made a drink by immersing the seeds in cold water. The Arabs developed the technique of preparing coffee by steeping the beans in hot or boiling water about 1,000 years ago.

Coffee is a small shrub native to east Africa, specifically the Ethiopian highlands. Its scientific name is *Coffea arabica*, even though it does not grow spontaneously in Arabia. Farmers also cultivate two other species of the same genus, C. *canephora*, robusta coffee, and C. *liberica*, Liberian coffee. Liberian coffee is not widely grown, but robusta coffee accounts for about 10 percent of all coffee crops. It is used primarily in the manufacture of instant coffees because of its higher caffeine content. Robusta is also more resistant to disease than *arabica* coffee. There are more than thirty species of wild coffees in Africa, Madagascar, and Asia. These species are potential gene pools for disease resistance.

As with tobacco, stories about the spread of coffee abound.[15] Muslim pilgrims supposedly drank coffee to keep themselves awake during long prayer sessions. At the beginning there was strong opposition to its consumption by orthodox priests, who considered coffee an intoxicating drink. It was prohibited by the Koran. The goods of coffee dealers were confiscated and the product was destroyed. In spite of these measures, coffee drinking spread until it became the national drink of Arabs. Travelers introduced the Arabian drink to Europe. In 1615 Venetian merchants imported coffee beans from Mocha. According to tradition, the Arabs were concerned about protecting their monopoly. Before allowing the beans to be exported they dipped them in hot water to kill them. At first Italian priests opposed the drinking of coffee, reasoning that since it came from the Muslim world it must be a product of the devil. Pope Clement VII settled the dispute in favor of coffee by "baptizing" it. Soon after, coffeehouses sprang up all over Italy and then spread to other parts of Europe. According to another legend Conopios, a disciple of Cyrill, the last patriarch of Constantinople, introduced coffee into England. After the Turks murdered Cyrill in 1637,

Drawing of a branch of the coffee shrub, seeds, and a fruit showing the two seeds, or beans. In the preparation of coffee beans, the pulp and the parchment skin have to be removed before roasting. A green bean is an unroasted bean, even though it is ripe.

Conopios settled in Oxford, where coffee became a favorite with students. They formed the Oxford Coffee Club, which later evolved into the Royal Society.[16] It is said that coffee was introduced into Austria after the siege of Vienna by the Turkish army of Kara Mustafa. According to the story, the retreating Turkish army left behind many sacks of coffee, which started the Viennese love affair with the drink. Coffee was introduced into France through the port of Marseilles, where coffeehouses appeared around 1671. Wine merchants in France fought against coffee, fearing it would replace wine.

Opposition to the use of coffee came from other quarters too. In 1777 Frederick the Great of Prussia issued the following declaration: "It is disgusting to note the increase in the quantity of coffee used by my subjects and the amount of money that goes out of the country in consequence. Everybody is using coffee. If possible it must be prevented.

My people must drink beer. His Majesty was brought up on beer, and so were his officers. Many battles were fought and won by soldiers nourished on beer; and the King does not believe that coffee-drinking soldiers can be depended upon to endure hardships or to beat their enemies in case of the occurrence of another war."[17] Similar sentiment is expressed in Johann Sebastian Bach's famous "Coffee Cantata": "You wretched child, you immoral creature, when am I to have my way? Remove that coffee from my sight!"

Coffee's opponents could not hold back the flood. Coffeehouses became centers for social, political, literary, and commercial life. Venerable institutions such as Lloyd's of London had their beginnings in coffeehouses. To this day people conduct transactions at Lloyd's across tables that resemble the pews in the original coffeehouse. Voltaire, Rousseau, Beaumarchand, and Diderot are among the many famous Frenchmen who transformed Parisian coffeehouses into centers for intellectual, political, and social intercourse.

Until the end of the seventeenth century all coffee came from Arabia. In 1616 a coffee plant was grown in Amsterdam's botanical gardens. Offspring were taken to Java, where the Dutch set up plantations. Plants were distributed to other botanical gardens, among them the Jardin des Plantes in Paris. The famous French botanist Antoine de Jussieu personally took care of that garden's plant, showing it to visitors with great fanfare. In 1723 a young French naval officer posted in Martinique conceived the idea of growing coffee there. After many difficulties he was able to transplant a single tree to Martinique, which gave rise to a thriving industry there and on other French islands in the Caribbean. From the Caribbean, plants were taken to Venezuela, Colombia, and French Guyana, and from French Guyana they diffused to Brazil. In time South America became the major coffee-growing area of the world, which it remains today.

The coffee plant is a shrub or small tropical tree that grows naturally in the understory of the forest. Though susceptible to frost, it does not grow well in the lowland tropics. Farmers prefer to grow it on lower-elevation mountain slopes in the tropics. They start plants from seed, and when the seedlings are about twenty centimeters tall they transplant them.

Coffee grows best in the forest understory. There yields are low, but the aromatic compounds that give coffee its flavor are concentrated. Coffee grown in shade is also ecologically more sustainable. Forest trees and their roots protect the soil from erosion, especially on mountain slopes. Litter from the trees serves as fertilizer, and the shade keeps weeds down. Shade plantations can be kept productive for twenty to thirty years and more.

Originally farmers grew all coffee in the shade. To increase production planters started experimenting by removing those trees that produced the greatest shade and keeping or planting those with feathery leaves. This added light in the understory and increased coffee yields somewhat, without substantially worsening soil erosion or weed density. If farmers removed all the trees, yields increased even more but serious ecological problems arose, the exposed slopes of tropical mountains being so prone to erosion. Coffee plants grown in the sun no longer obtained the benefits of natural fertilization from trees above, and weeds invaded coffee plantations. The quality of coffee decreased but yields increased dramatically. Sun-grown coffee, unless fertilized heavily, exhausts the soil in less than twenty years.

The first coffee plants arrived in Brazil in 1727 from French Guyana, supposedly smuggled by a lieutenant colonel of the Brazilian army, Francisco de Melho Palheta. It was said that he obtained a cutting hidden in a bouquet of flowers presented to him publicly by the governor's wife at a banquet. Coffee was first grown commercially in Brazil in the state of Espíritu Santo, north of Rio de Janerio.[18] By the beginning of the nineteenth century, Europe and North America increased in population and wealth, creating more demand for coffee. During the eighteenth century demand had been supplied by the plantations in the East Indies, primarily Ceylon (today Sri Lanka). But the coffee blight native to Africa invaded that region in the early 1800s, just when Brazil was starting its production. The combination of more world demand and less supply from the traditional source created a tremendous call for South American coffee. A boom economy took hold not only in Brazil but in Colombia and Venezuela as well.[19] Coffee growing transformed sleepy haciendas into bustling enterprises run with slave labor. Forests were cut down and trees burned. Initially the ashes provided fertilization for the young seedlings. After about two decades soil exhaustion and erosion combined to reduce yields. Farmers then cleared new land and repeated the process. The abandoned coffee plantations slowly reverted to forest. Coffee became a nomad crop that moved from the state of Espíritu Santo to the hinterland of the state of Rio de Janeiro in the middle of the nineteenth century, to the state of Minas Gerais in the latter part of the century, then to the state of São Paulo.

Brazil prohibited slavery in 1888. Traditional planters, their soil exhausted, their plantations aged, and their prevailing labor source gone, went bankrupt. The newer farmers of São Paulo adjusted to the situation by employing cheap immigrant labor. Portugal, Italy, and Japan were the primary labor sources. Coffee started in the eastern part of the state and slowly and inexorably moved west as soils became exhausted. Immigrants who had been able to accumulate some savings bought the exhausted land. With their labor and effort they restored the land and developed a more diversified farm economy. Still, after World War I Brazil's economy became totally dependent on this one crop. While 70 percent of export earnings came from coffee, the benefits to the country were limited. A small elite controlled the coffee-export business, and the number of producers was likewise small, being dominated by large landowners. In order not to adversely affect the export of coffee, free trade prevailed and Brazil imported most manufactures. It became a prime example of a dependent economy.

When depression hit the United States and Europe in the 1930s, demand for coffee plummeted. Faced with economic hardship, Europeans and North Americans reduced their coffee consumption. Many countries, especially in Europe, imposed import restrictions on coffee. Overproduction led to a dramatic drop in prices and Brazil plunged into a severe economic depression. Reduced export earnings forced drastic import restrictions, Brazil was forced to default on its foreign debt, and to reduce supply and improve prices, coffee was burned as a fuel. The situation did not improve until after World War II was over and demand picked up.

During the Depression farmers abandoned many aging coffee plantations. After the war coffee continued its movement south into the state of Paraná. Climatic conditions stopped it there. While frost occurs only occasionally in São Paulo, it is common in Paraná. In the 1960s when frost twice destroyed the coffee plantations of the state, an era in Brazilian agriculture came to an end. Coffee could no longer be grown as a nomad crop on virgin land, because suitable sites no longer were available. From then on it had to be grown as a sustainable crop. Farmers applied fertilizer and took other measures to reduce soil erosion. The crop moved back north. Today coffee is grown in several states of Brazil, primarily in the state of Minas Gerais.

Coffee is no longer the mainstay of the Brazilian economy, although Brazil is still the largest coffee producer in the world. Today the crop accounts for only 7 percent of the country's export earnings. And while coffee is still primarily a sun-grown crop, efforts are being made to improve its quality and price.

Coffee, sugar, wheat, and grapes are examples of crops that the Old World gave the New; potatoes, tomatoes, maize, and tobacco traveled the other way. Because of the long isolation between the two hemispheres, and improved communications in Columbus's time, the Columbian exchange had dramatic effects. It transformed diets as well as economies throughout the world.

Consider a typical Italian dish, spaghetti with tomato sauce. Pasta is made from wheat from the Middle East; tomato sauce is made from tomatoes and sometimes red pepper (South America), onions (Europe and Asia), and garlic, oregano, basil, and rosemary (the Mediterranean). Or consider a traditional American Thanksgiving dinner. The turkey is, of course, native, but the stuffing contains wheat, celery, onions, and herbs, all from across the Atlantic. Potatoes, maize, and squash are all from South America or Mesoamerica, but the table might be laid with some Old World vegetables such as peas or lettuce. Cranberries are North American, and pumpkins are from the New World. However, to make a pumpkin pie you need the nonnative animal products milk and eggs, as well as sugar (Oceania or Europe, depending on whether it comes from the sugarcane or the sugar beet), and the Asian spices nutmeg, cinnamon, and ginger. Even our ordinary national meals are international celebrations.

CHAPTER

A New Kind of Farm

ONE OF THE serious environmental problems we face today is the loss of landscapes, species, and genetic diversity, the so-called biodiversity of the planet. This is the direct result of the transformation of natural landscapes by human action, primarily though not exclusively for agricultural purposes. Biodiversity is lost when tens, hundreds, thousands of farmers clear land to increase their arable surface, when loggers clear forests to provide lumber for the construction of houses and furniture, when city dwellers and suburbanites build houses, factories, schools, and roads. The accumulated actions of all these people lead to the extinction of species and the irretrievable loss of their genetic makeup. The process has been going on since humans invented agriculture 10,000 years ago. It accelerated markedly in the seventeenth and eighteenth centuries with the industrial and agricultural revolutions and the advent of modern farming.

The industrial revolution spanned the second half of the eighteenth and the first half of the nineteenth century in western Europe. It saw the change from a predominantly rural economy to an industrial economy based on the use of steam and fossil fuels. A period of demographic growth in Europe, it was marked by a massive migration of population from countryside to city.

We associate the industrial revolution with the invention of the steam engine and its use in mining, ships, and especially railroads, with the accumulation of wealth and the emergence of a strong middle class, and with the exploitation of labor, including the child labor so vividly described by Charles Dickens and others. None of these changes could have taken place without a great improvement in agricultural productivity. The enclosure movement made possible innovative agronomic techniques and uses of crops. Continuing experimentation resulted in overall increases in productivity. And commercial cereal agriculture expanded to areas outside Europe, in particular North America, Argentina, and Australia.

A major industry to emerge from the industrial revolution was the manufacture of cotton cloth in eighteenth-century England. A hundred years later the availability of vulcanized rubber for tires sparked the development of the car industry. Thus two plant species provided important raw materials that supported an unprecedented level of industrial expansion. But it was not industrial crops alone that made the industrial revolution. Foodstuffs to feed increasing urban populations played an essential role, and the world experienced a substantial increase in agricultural surplus. New agricultural technologies contributed significantly to the production of this surplus, as did the opening of new land to agriculture in the Americas and the introduction in Europe of the potato and other American plants.

A rapid increase in the population of Europe took place in the eighteenth century. The population of England and Wales went from about 5.5 million at the beginning of the century to 9 million at its end.[1]

France, which had about 20 million inhabitants in 1700, ended the century with 26 million. Other West European countries such as the Netherlands and Spain experienced similar growth, though it is harder to obtain exact figures for them. The increase took place in the midst of rising living standards and without the general famine that had characterized previous centuries, even though local crop failures continued to occur. The plague that had affected so much of Europe intermittently since the fourteenth century virtually disappeared.

Nobody knows exactly why the population expanded. Some attribute it to an increased food supply, others to improved sanitary conditions or the movement of people to the cities.[2] Those who believe that an increased food supply was responsible point to the expansion of potato growing. The potato, if not the cause itself, certainly made it possible to feed an increasing number of people.

In any case, to feed the expanding population agricultural production had to increase. In the fourteenth century, the only possible response to the needs of a growing population had been to augment the cultivated surface; in the eighteenth century, the response was also to increase output of existing surfaces. This was possible because of advances in agronomy. Another difference was that although the population as a whole was growing, the proportion that produced its own food was declining. The nonfarming population had more purchasing power, which meant an increasing demand for higher-value agricultural products such as white bread. Demand encouraged regional specialization and promoted rural division of labor, especially among farmers situated near the emerging industrial centers. Expanded demand sent up prices and rents from farm leases and raised the value of agricultural land, a further sign of prosperity.

Other sources of demand for agricultural products were emerging industries. First among them was the weaving and clothing industry. At first, both in England and France, weavers used mainly wool. England, France, and Spain, the principal wool producers at the time, could hardly fill the demand. Eventually abundant cotton from India and later from North America replaced wool. Tallow for candles and for grease to lubricate the gears in machinery, hides for leather to make belts and harnesses, and bones for glue were also in great demand. The brewing industry used more barley than ever before. And flax was required for the manufacture of linen.

In response, in the seventeenth century western Europe adopted the crop-rotation practices that Flemish and Dutch farmers had pioneered in the previous century. Farmers planted cereal, still the principal source of food, in better soil with more fertilizer and put more effort into harvesting efficiently. This resulted in a better-quality flour that made a superior bread when baked. White bread took its place at the top of the list of desirable foods, causing a decline in the demand for brown and black bread. Farmers planted fallow lands with legumes such as beans, peas, clover, and alfalfa. In France, vineyards expanded and soon the country was producing an excess of wine. Efforts were made, especially by the French nobility, to improve the quality of wine, and a taste for good vintages developed among the well-to-do. The quality of ordinary wine improved as well.

Landed elites ruled both England and France. The greater value of land and therefore the social and political desirability of owning it encouraged landowners to use better agronomic methods to recoup their investments. This required control over the land, increasing the pressure for enclosure and the extinction of medieval common rights. Once an estate was enclosed it could be leased for considerable periods. The greater revenue of the land brought more investment, and progressive farmers sought new ways to improve productivity and profitability. New agricultural machinery was introduced, such as iron plows and seed drills. The horse replaced the ox as the source of power.

Not everybody in the countryside profited equally from these changes. Most fortunate were the landowners, who benefited from the higher value of their properties. Next in line were the large and progressive tenant farmers who leased the best land. Last came the poor landless population, who had only their labor to offer. Wages in a time of expanding population rose much less than the price of food or rent. Historians calculate that between the start of the eighteenth century and the French Revolution, landed income rose in France by almost 100 percent, prices by about 60 percent, and wages by 25 to 30 percent.[3]

Traditionally it was thought that agricultural change paralleled the development of industry. Historians feel today that agrarian change was more gradual, preceded by a long period of development stretching back to the sixteenth century and the beginning of the enclosure movement. The principal agricultural innovations during the industrial period were the introduction of new fodder crops, primarily turnips and clover, and a reduction in bare fallows; potato growing; the systematic use of fertilizer, including chemical fertilizers and guano, the latter imported from Peru; higher-yield varieties; the draining of heavy fields; and efficient, easily repairable machinery.

The eighteenth century saw a considerable increase in agricultural production, especially of grain, milk, beef, and wool. Both yields and cultivated surface expanded. Precise figures are hard to come by, but historians estimate that wheat yields went from twenty to twenty-two bushels per acre in 1700 to twenty-four to twenty-seven bushels per acre by 1850, a 25 percent rise in yield.[4] Total output, however, doubled over the period because cultivated area grew by as much as 50 percent. Compared to an increase of about 3 percent per annum in agricultural production worldwide over the last twenty-five years, a doubling in 150 years is modest. Yet it is significant when compared with the almost flat production of the preceding 300 years. Agriculture was able to feed an expanding population and to provide increased quantities of wood, leather, tallow, and wool to emerging industries. Many more horses were employed to pull equipment on farms and in industry, as well as for private carriages. Historians calculate that there were 3.5 million horses in England at the end of the nineteenth century.[5] Since each horse consumed in a year the fodder of about 5 acres, some 15 million acres were required to maintain them, the equivalent of 400 million bushels if the land had been planted to wheat!

England is a model for the industrial revolution because it took place there early and comparatively quickly, over a period of a century. The industry most revolutionized was the textile industry, especially cotton processing and weaving. One impetus for the enclosure movement that had transformed traditional agriculture in sixteenth-century England was the demand for wool. Sheep farming employed less labor than grain farming, leading to rural unemployment. As enclosure accelerated in the seventeenth century, England shifted from exporting wool to exporting undyed cloth woven largely by local villagers.

Cloth was woven almost everywhere in England. Most of it was for local consumption. By the beginning of the eighteenth century, certain regions had specialized in the production of fine cloth for export. The finest cloth came from the west of England and from East Anglia. Initially wool and linen were the fibers used in the manufacture of fabric. Starting in the seventeenth century, cotton imported from India was added to the list of fibers. Cotton manufacture was initially centered in Lancashire.

Because the traditional tools for cloth making—spinning wheels and looms—were simple, easy to acquire, and operated entirely by hand, there was no need to concentrate them in one place. Therefore most cloth was woven locally according to what persons in the industry called putting out. The system was highly flexible. In some instances the weaver was an independent craftsman buying raw materials on credit and selling the finished product to a local merchant. Other weavers worked for an employer who provided raw materials and paid a fixed price per piece. Many weavers combined weaving with some agriculture. Sometimes weavers worked alone; in other cases they employed two or three apprentices.

With the industrial revolution came new, highly productive machines for spinning yarn and weaving cloth. The new cloth-making machinery represented a sizable investment that individual weavers could not afford. It also required a source of power other than human, either water or steam. Consequently specialized factories located in cities came into being. The greater efficiency of these plants brought ruin to the rural cottage industry.

The growth of employment in textile mills led to slow increases in the buying power of city populations and fed the demand for food and other agricultural produce—wheat for bread, barley for malting, and dairy products, beef, mutton, and poultry. This was particularly true of London. Much of England was employed in supplying that city. Cheshire, Wiltshire, and Suffolk counties provided it with dairy products, Leicestershire and Lincolnshire with bread and cattle. East Kent and Worcestershire specialized in orchard products and hops. Much of the Midlands specialized in growing wheat and barley; the northern countries specialized in oats.[6]

Before the eighteenth century, cities with more than 10,000 inhabitants had been few. Their food came from small areas surrounding the city. Because transportation costs were high, grain was rarely hauled overland more than twenty miles. This changed with the introduction of steam power and railroads, which drastically reduced the cost of land and water transport.

The need for more foodstuffs, the growth of the population, and the movement of the rural population to the city required an intensification of farming to keep up with demand. In other words, the farming sector needed to obtain greater yields per unit surface and especially per unit labor. Farmers needed better farm implements and better soil-management techniques. Tools that until then had been manufactured by local craftsmen began to be produced industrially, since the new factories could generate a great diversity of iron plows, seed drills, reapers, and threshing machines of better quality and at a lower price than local craftsmen. New theories on plant nutrition led to the development of the chemical-fertilizer industry. Farms became more commercialized and dependent on outside inputs. The horse, and eventually the tractor, replaced oxen as the source of power to move machines. Subsistence farming, which predominated in the early seventeenth century, had virtually disappeared in England by the middle of the nineteenth.

The introduction of railways and steamships revolutionized transportation by dramatically lowering costs. It now became economical to import foodstuffs from faraway places. The newly opened lands of Argentina, Canada, and the United States began to undersell European grain producers in the second half of the nineteenth century. Vegetable oils became important, not only for cooking but also as lubricant, for paint, and for other industrial uses. Linseed from Argentina and coconut and palm oil from the tropics increasingly supplied the English market, while English textiles penetrated the remotest parts of South America. And the need for efficient belts to drive machinery created a demand for a new tropical product, rubber. But no crop was more important to the industrial revolution in England than cotton.

The earliest known fragments of cotton cloth come from India and are approximately 5,000 years old. These exhibit good craftsmanship, sug-

The vegetable lamb, after Sir John Mandeville.

gesting that cotton weaving in India must be older still. We have cotton fabric made by twining from northern Peru dated about 4500 BP and woven fabric from Peru dated a little later.

Cotton arrived late in the Mediterranean, preceded by legends. Herodotus said of India, "Certain trees bear for their fruit fleeces surpassing those of sheep in beauty and excellence, and the natives clothe themselves in cloths made therefrom."[7] In Europe a legend sprang up of a "vegetable lamb," a combination of animal and plant. According to one version, cotton was a tree with fruits that when ripe opened and yielded lambs. Another version had a tree trunk topped by a sheep that grazed around the tree.

Cotton fibers are single-celled seed hairs that grow from the seed coat. The cotton seed produces two types of hairs. Long hairs, known as lint, are 20 to 50 millimeters long according to species and are easily pulled free of the seed. Much shorter hairs a few millimeters long, called linters or fuzz, stay attached to the seed. The amount of lint hair on the seed is such that at maturity it breaks open the cotton fruit, or boll. The original function of the hairs was probably to attach the seed onto passing animals and thereby disseminate the species. The lint hairs in some cotton species can be tightly twisted, which is what makes it possible to spin cotton.

Cotton seeds are good sources of oil. The oilcake left when the oil has been pressed out is high in protein and good as animal feed. The linters are used for making rayon, which is largely cellulose. Thus farmers can use all aspects of the cotton seed.

The botanical history of cotton is interesting and puzzling. There are many wild species and varieties of cotton in tropical and subtropical areas of Africa, Asia, Australia, and the Americas. Wild species grow as shrubs, small trees, or perennial herbs. But there are only four linted species, which may have been domesticated independently. The Old World species are all diploid, with twenty-six pairs of chromosomes. Either the New World species are diploid, with twenty-six pairs of chromosomes, and lacking twisted lint, or they are tetraploid, with fifty-two chromosome pairs, and produce lint. Closer examination shows that the chromosomes of Old World diploids are larger than those of New World diploids. The two New World tetraploid species, sea-island, pima, or Egyptian cotton and upland cotton, have twenty-six pairs of each kind. Scientists conclude that the South American tetraploid species resulted from a cross between a New World and an Old World species in prehistoric times before domestication. The cross probably took place in the lowland Amazon, where several diploid South American species grow.

The history of Old World cotton is complex. The ancestral species, *Gossypium herbaceum*, was originally an inhabitant of southwestern Africa. From there it spread across Africa to Mozambique and then to Arabia, where it was probably first domesticated. From Arabia it was taken to India. It is believed that a mutation there gave rise to the fourth linted species, G. *arboreum*.

In spite of its name, G. *herbaceum* is a shrub, while G. *arboreum* is a shrub or small tree. The two American species are small shrubs or pe-

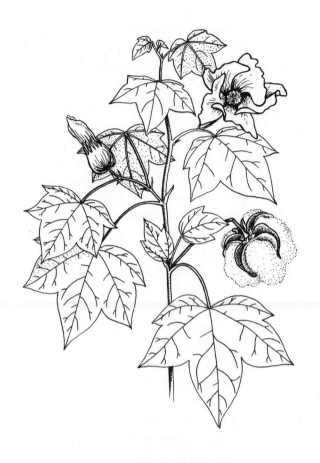

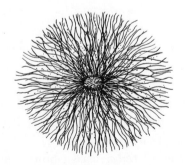

Upland cotton and cotton seed showing the seed hairs used in weaving.

rennial herbs. In temperate areas cotton, which cannot stand frost, behaves as an annual and farmers must replant it each year. The New World species have longer fibers and are the most widely cultivated today.

During the eighteenth century, cotton fabric imported to Europe from India became popular. Until then, because of the high cost of processing the fiber, cotton was a luxury cloth. The task of manually separating the lint from the small seeds was time consuming. After ginning, as this task is called, the fiber had to be cleaned, baled, carded, and spun. Historians have calculated that before the industrial revolution it took twelve to fourteen man-days to produce a pound of spun cotton thread, while it took only one to two man-days for wool, two to five for linen, and about six for silk.[8]

This situation was to change entirely with the invention of the cotton gin and the mechanization of spinning and weaving that marked the beginning of the industrial revolution. These developments changed the economics of cotton-cloth production. Cotton textiles were well suited for mechanization, for the large number of man-hours needed to prepare them meant that improvements in efficiency could be profitable, and cotton fibers were not too brittle or soft for handling by machine.

During the industrial revolution yarn quality improved, the production of cotton cloth dramatically increased, and prices fell, all of which spurred demand. Inventions such as the fly shuttle, spinning jenny, spinning mule, and power loom made possible production on a scale unknown before. Their effect in England was immense. In the twenty years from 1765 to 1784, the quantity of cotton spun there rose from less than 0.5 million pounds to over 16 million. By 1850, 60 percent of England's exports was cotton cloth, and 25 percent of the world's production was woven in England. In 1860 England imported over 1 billion pounds of cotton.

In 1793 the American Eli Whitney invented his cotton gin. Ginning had been the bottleneck in cotton production since it required so much hand labor. Whitney's cotton gin could remove seeds from cotton fifty times faster than by hand. The American cotton gin was not the first. In India the churka gin utilized rollers to squeeze out the seeds.

But the churka gin was not effective with the long-fibered upland cotton planted in the United States.

Until the invention of the Whitney gin, India had been the major supplier of cotton to England. The United States had always had the advantage of location, which in the days of sailing ships and high transportation costs was significant, but higher labor costs made American cotton uncompetitive. This changed with the cotton gin. Its appearance just at the time when the demand for raw cotton was accelerating in England gave a huge impetus to cotton growing in the American South. Cotton acreage increased, and cotton exports went from less than 10 million pounds in 1800 to more than 2 billion pounds in 1850. The increase in cotton growing had an unfortunate side effect in that it gave a great boost to the institution of slavery. Until the end of the eighteenth century, the Southern economy had rested primarily on tobacco, for which the market was declining. Slavery was not especially economical. When the gin enabled a person to remove the seeds from 50 pounds of cotton a day, as opposed to only 1 or 2 pounds by hand, the value of slaves went up. As farmers planted more and more land with cotton, they needed more slaves. The South became dependent on cotton for its prosperity and on slavery to produce its cotton. The dominance of cotton was thus one of the principal causes of the Civil War in the United States.

Cotton also affected the landscape of the American South. Land in colonial days had been cheap, and often people would try to get short-term benefits from the soil, then abandon it and move on. This was particularly true for those growing cotton, which is hard on the land. Early cotton growing was concentrated in the Carolinas and Georgia. As land wore out and demand grew, cotton moved west into Tennessee, Alabama, Mississippi, and Louisiana. After the Civil War Texas became a big producer, and later the Western states of California, Arizona, and New Mexico.

Along with the sugar beet, rubber is the most important crop plant domesticated in historical times. The rubber tree (*Hevea brasiliensis*) that is native to the Amazon rainforest supplies most commercial rubber. It is not the only rubber-producing plant. There are several tropical trees that produce small amounts, and a relative of the common dandelion as well as the guayule shrub of the Southwestern United States and northern Mexico produce rubber. But the rubber tree (not to be confused with the houseplant of that name, which is really a fig) produces more rubber that is easier to extract than any other source. It has been, and continues to be, the principal source of natural rubber.

Rubber molecules are contained in solution in special cells called latex cells, found in the bark of the rubber tree. When tappers cut the bark, the content of the latex cells oozes out for a number of hours until the wound heals. When exposed to the air the liquid congeals and has an elastic quality. The Indians of the Amazon discovered the plant and the properties of its latex and with it fashioned solid rubber balls. However, natural untreated rubber is unstable, and for centuries the product remained mostly a curiosity. Its only use was for waterproofing

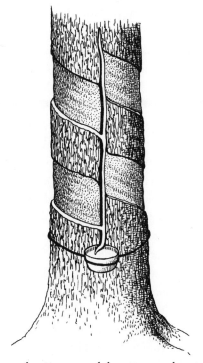

Stem of rubber tree showing cuts made by tappers, and a cup into which the rubber flows.

boots or other articles, until the last century, when a set of circumstances arose that would revolutionize life-styles the world over.

The first development was the invention of vulcanization by the American Charles Goodyear in 1839. When rubber is mixed with sulfur in the right proportions and heated, it stabilizes, forming a durable product. The second development was industry's new mechanized equipment. Vulcanized rubber proved to be a much better material than leather for gaskets and belts. Soon industrialists found other uses for the product: tubing, buffers between parts, and bicycle tires. Demand for the product increased dramatically. Great Britain imported 211 kilograms of raw rubber in 1830, 10,000 in 1857, and 58,710 in 1874.

The increasing demand for rubber produced an economic boom in Brazil. Rubber trees grow dispersed in the rainforest. Tappers, called *seringueiros* in Brazil (from *seringa*, Portuguese for rubber), had exclusive rights to one or several trails in the forest, which they cleared themselves. In most cases they kept the location of the trails secret. A trail had anywhere from 60 to 150 trees. *Seringueiros* walked their trails twice every other day during the tapping season. On an afternoon visit, they gathered latex that had oozed out of V-shaped cuts in the bark made in the morning and that had collected in small cups under the wound. They carried the liquid to a camp and slowly poured it onto a long pole placed horizontally above a smoking and smoldering fire. The rubber congealed into a large ball a meter or more in diameter, which would be shipped to a trading center such as Manaus or Belem.

The tapping season lasted for about half of the year and was restricted to the dry season. A tapper produced between 200 and 800 kilograms of rubber in a season, with an average of 500. Life for the *seringueiros* was hard. During tapping season they lived in the jungle, exposed to the dangers of accidents, parasites, and diseases. Most worked in the jungle only a few years, hoping to earn enough working capital to return to the city. Traders called *aviadors*, or forwarders, provided needed supplies to *seringueiros* in exchange for the exclusive right to the rubber. These middlemen sold the rubber to exporters. Profits were high. The rubber boom transformed the sleepy Amazonian town of Manaus at the intersection of the Negro and Amazon rivers into a bustling metropolis. The "rubber barons" built sumptuous living quarters, and the city of Manaus built a French-style opera house in which European companies performed. The rubber boom lasted from the middle of the nineteenth century until the end of World War I.

The manner of harvesting rubber in Brazil was uneconomic and created unnecessary exploitation of tappers. Domestication of the rubber tree on plantations was the logical solution. But there was little incentive to change the system as long as demand outstripped supply and prices remained high, and only minimal cultivation took place.

In the meantime the British, the principal consumers of rubber, started planning to create rubber plantations in the Far East. The moving force behind this enterprise was Clement R. Markham, at the time a functionary of the India Office. He persuaded the Foreign Office to telegraph the British consul in Belem asking him to obtain rubber tree seeds. Henry A. Wickham, a young English adventurer and drifter, was given the task of gathering the seeds. This he did, and in 1876 he forwarded some 70,000 seeds to Kew Gardens in Richmond, on the outskirts of London. After they germinated, 1,919 seedlings were sent to the Royal Botanical Gardens at Peradeniya, Sri Lanka. Seedlings were also sent to Singapore and other tropical British colonies. The Sri Lanka plants started bearing seeds in 1882, from which seedlings were produced to start plantations in Sri Lanka, Burma, and India. In 1881 trees were tapped and the rubber was deemed satisfactory. It was in Malaya, however, that rubber planting had its greatest success, and in 1898 the first commercial sale of rubber was registered there. Rubber from the Far East quickly increased in volume and by 1913 overtook Brazilian rubber production. After World War I, the volume of cheap plantation rubber from the Orient resulted in a dramatic drop in price. Rubber had been successfully domesticated, and with it the Brazilian boom in wild rubber ended.

Brazilians tried to establish their own plantations but soon discovered a devastating leaf blight that killed or severely restricted production. Their isolation from other rubber trees protects individual trees in the forest from blight, but once the blight invades a plantation it quickly spreads. In World War II, when the Japanese conquered Malaya and took possession of rubber plantations, the United States and

Brazilian governments engaged in a deliberate effort to conquer the disease, but they failed. Had the British imported seedlings directly from Brazil to the Far East they might have imported the blight, in which case the story of rubber and all its industrial uses might have been radically different.

Cheap plantation rubber became available at the same time as mass-produced cars. Demand for rubber skyrocketed. Eventually chemists in Germany and later in the United States learned to synthesize rubber. Today a large part of commercial rubber is synthetically produced.

The other major crop domesticated during the industrial revolution was the sugar beet. Its story, though different from rubber's, also illustrates the inventiveness and the entrepreneurial spirit of the time.

Unknown to most Europeans until recently was that while nonnative sugarcane was making its impact felt on the social and economic life of Europe, a sugar-producing plant grew in their midst. This was the sugar beet, a variety of the common beet used primarily as forage. The beet grows wild in the Caucasus and around the Caspian Sea. The first mention of its cultivation was by Aristophanes around 425 B.C. It was grown in ancient Rome and throughout the Middle Ages.

In 1747 Andreas Sigmund Marggraf, a German chemist, found that some varieties of fodder beets contained sucrose, although the concentrations were low. Farmers around Magdeburg, Germany, used the beets to extract a sweet syrup, a practice that may have preserved the sweet varieties of the plant. One of Marggraf's students, F. C. Achard, recognized the commercial potential of his teacher's discovery. Achard started a series of studies to develop ways to extract the sugar from the beet. With the proceeds of a grant obtained from the king of Prussia, Frederick Wilhelm III, Achard installed a pilot plant in Cunern, Silesia. Achard also recognized the importance of increasing yields by selection. Unfortunately the yields he obtained were low and he died a disappointed man, not realizing that his idea would soon take root throughout Europe. Another sugar beet pioneer was Freiherr von Kopy, who on hearing of Achard's experiments began his own factory and beet-breeding program. From his work derives the white Silesian beet, the "mother stock of all the sugar beets of the world."[9]

From here the action moves to France and the time of the Napoleonic wars. To thwart the sugar trade controlled by the British, Napoleon subsidized factories for the extraction of sugar from beets and ordered the cultivation of the white Silesian beet. Yields were still low. Philippe Andre de Vilmorin and his son Louis began an ambitious breeding program to improve the productivity and sugar content of beets. Over a period of fifty years they doubled the sugar content of beets, making them competitive with sugarcane. Meanwhile, the French chemist Claude Louis Comte Berthollet applied himself to the problem of extraction and developed efficient methods.

The sugar beet was to transform the sugar industry. Now there were two species, one tropical and the other temperate, that could produce sugar relatively cheaply and efficiently. This added competition pressured the sugarcane industry to improve its performance, which it did on plantations in the New World.

As we have seen, the benefits of the industrial revolution were unevenly divided. Rents and prices rose more steeply than wages, and poor tenant farmers who lost their privileges under the old common-field system did not find respite in the new system of leaseholds. To obtain the cash for rent, smallholders had to sell much of their grain production and rely on the higher-yield potato newly introduced from South America to meet their own food needs. The danger of dependence on one crop is nowhere better illustrated than in Ireland.

Although precise figures are difficult to obtain, historians suggest that the population in Ireland grew 172 percent between 1770 and 1841.[10] Enlarged food supplies resulting in part from potato growing made it possible to feed the greater numbers. Ireland was the first country of northern Europe where the potato had become firmly established. Although the record is scanty, historians agree that by the end of the seventeenth century the potato had become the staple of the impoverished Irish farmer.[11] It did not displace grain. Barley, oats, and some wheat were cash crops grown to pay the rent on the land. Potatoes, which would grow in heavier and marshier soils, where farmers could not plant grain, provided food for tenant farmers.

Ireland had been in a perpetual state of rebellion since the English first invaded in the twelfth century. Periods of apparent calm alternated with periods of open warfare. The seventeenth century was particularly bloody. Both sides, but especially the English, practiced a type of warfare invented shortly after the adoption of agriculture and perfected by the Greeks: the systematic destruction of agricultural fields. "At Clogher, in the south of Tyronne, we stayed to destroy the corn; we burned the country for twenty-four miles compass and we found by experience that now was the time of the year [end of September] to do the rebel most hurt," reported the English deputy Sir Henry Sidney in 1567.[12] In the sixteenth century the Irish peasantry were mostly cattle herders rather than farmers. During the wars large quantities of livestock were killed. Cattle herds being more difficult to replace than agricultural crops, the rural population increasingly turned to farming to sustain themselves.

The Penal Laws, dating from 1695, were a reaction of the English to the support the Irish had given the Stuart king James II after the Protestant William of Orange ascended the throne of England. The Penal Laws barred Catholics from the army and the navy, the law, commerce, and all civic activities. No Catholic could buy land, and Catholic estates were to be divided among the sons upon the death of the owner. Primogeniture was to be invoked only when the oldest son converted to Protestantism. These laws were especially hard on the peasantry.

This social situation set the stage for the adoption of the potato. Potatoes were more difficult to destroy than grain in the field and harder to burn than stored grain. Potatoes produced more bulk, could be fed to both humans and livestock, and were superbly adapted to the damp, cold climate of Ireland. They were introduced into Ireland in the first half of the seventeenth century (there is no exact record when or where) and slowly spread. By the end of the century most Irish had adopted them, and for the next 150 years potatoes were the staple that fed the peasantry.

By this time in Irish history, farmers had brought under the plow all available land. As the population grew, farmers subdivided plots among more and more tenants until land that had once been leased to one family supported four or five, sometimes even ten, times that number. The resulting competition for land increased the price of rents, which were as much as 100 percent higher than in England.

Absentee landlords, who rarely visited their holdings, owned most of the land. The usual practice was for the owner to rent his land on a long-term lease and at a fixed price to a middleman. The middleman in turn rented parcels on short-term leases to tenant farmers. This system gave landowners a fixed and secure income and left the management problems to often ruthless middlemen. The latter leased the land to the highest bidder. Since tenant farmers were for the most part without capital, they paid their rent after the harvest. Being in arrears most of the time, they lost the limited privileges they would otherwise have had. Sometimes because of weather the crop failed. This happened repeatedly in different districts in the 150 years preceding the famine. When tenants could not pay their rent they were evicted and became beggars or starved to death.

These were the conditions before the Irish potato famine. The famine of 1845 and subsequent years resulted from a conjunction of forces. The agent was the potato-blight fungus *Phytophthora infestans*, introduced from America, where it had been recognized as early as 1843. The year 1845 started dry and sunny but in mid-summer turned wet and damp, which favored the growth and spread of the blight. Healthy potato fields turned almost overnight into rotting heaps. People who tried to eat the infected tubers got seriously sick. Unlike previous crop failures, which had been local, this one was universal. Nor was it restricted to Ireland, spreading to England and the continent as well. Extreme famine followed, accompanied by the horrors that have been recounted a thousand times. The potato famine resulted in a massive emigration of peasants and renewed the determination of the Irish to regain their independence. At the center of this social upheaval was the country's dependence on one crop.

Another potentially disastrous plant pest was the phylloxera that almost ruined the entire European wine industry in the second half of the nineteenth century. In 1860 French vineyards were attacked by *Phylloxera vastatrix*, commonly called a root aphid although it is really a bug. The pest, which attacks roots, had been accidentally introduced in the 1850s with a shipment of American grapes and quickly spread.

North America had several species of wild grapes. The native inhabitants consumed wild fox grape, riverbank grape, and scuppernong. Early European settlers found that although good to eat, these grapes were not suited for making wine. Attempts to introduce the European grape into eastern North America failed because of cold winters and, unknown to its promoters at the time, the existence of phylloxera. The Spaniards in California were more successful. In 1784 Father Junípero Serra successfully introduced European grapes and wine making into the missions he set up. Before that, the Spaniards had successfully introduced wine grapes into Peru, Chile, and Argentina.

In 1852, E. W. Bull of Concord, Massachusetts, produced a new grape that he named Concord. This grape was far superior to anything grown until then in the United States, both as a fruit and for making wine. To this day, most grape juice is made from the Concord grape. It may be a mutation of the fox grape, or more likely a hybrid of the fox grape and the European grape that Bull was cultivating in his garden.

After the introduction of phylloxera into Europe, agronomists discovered that American rootstocks were resistant to the insect. The problem could be solved by grafting European wine grapes onto the resistant American rootstocks. Thus vintners undertook the costly process, using mostly American fox grape as rootstock. Today only isolated areas such as in Greece grow European wine grapes on native rootstocks.

The settlers who came to America in the seventeenth century remembered the open-field farming tradition of Europe, and except perhaps for the Puritans of New England, they had not self-sufficiency but commercial farming as one of their purposes in coming to the new land. For the most part they were greatly disappointed. During the early years death from starvation, disease, and Indian wars was so high that simple survival became the goal. This situation lasted for almost two hundred years. It was only in the nineteenth century, as a result of the demand of the industrial revolution for more food and fiber, that American farming began to expand into the highly productive system it is today.

Remote markets, poor land communications, and low production conspired against the early settlers. European production techniques and crops did not adapt well to the new environment, and the colonists had to devise new agricultural practices. Two American crops, maize and tobacco, became the mainstays of North American agriculture in the early days. The first was for a long time the chief food staple, the second the principal cash crop.[13]

Two elements, land and labor, were key factors during the main expansion phase of American agriculture. There was plenty of the former and a shortage of the latter. Abundant cheap land meant that farmers could move westward whenever the productivity of their current tracts decreased, or debts overwhelmed them, or the spirit moved them. As in other times and places, abundant land also meant that agriculture would be extensive rather than intensive, and that land would be abused.

Land speculation characterized the expansion of agricultural settlements in the nineteenth century. The hope of pioneers was not so much that they would make money by farming but that their land would skyrocket in value. So farmers were not always interested in becoming productive. Each pioneer family acquired as much land as it possibly could. This they tried to develop, clearing and fencing some fields, building some primitive housing, and improving roads. They waited until some land-hungry family with ready cash appeared, then moved on to repeat the process farther west. The second wave of migrants would settle and become commercial farmers.

The process of pioneering resulted in exploitative, unproductive farming. Pioneers planted crop after crop of maize or wheat on the same fields, adopting superficial and minimal tillage. They allowed cattle to run loose in the woods, producing little in the way of meat or milk. And in the long run land speculation usually proved unproductive.

Abraham Lincoln's father, Thomas, was such a farmer. In 1803, at the age of nineteen, he bought a 238-acre tract near Mill Creek, Kentucky.[14] In 1808 he added another 348 acres of unimproved land nearby. It was here that Abraham Lincoln was born. The soil was poor and the yields were bad. Partly because of this, and partly because of problems with the titles to the land, the Lincolns moved again in 1816, this time to a new tract of land in present Spencer County, Indiana. Here he bought 80 acres and settled down to open up the frontier. After another fourteen years of work he sold his farm at a loss and moved on to Illinois. Such was the life of the pioneer.

In the middle of the nineteenth century, and especially after the introduction of the railroad, American farmers had better access to markets, and this allowed them to become commercial farmers. The increasing American population and European need for grain provided ready markets. Land continued to be plentiful and cheap while labor and capital remained scarce and expensive. Farmers tried to make a profit by reducing the costs of production. Consequently they combined as much as they could of the cheap ingredient, land, with limited amounts of the expensive inputs, labor and capital. The result was extensive farming with low yields per acre. To raise production, a farmer simply increased the surface under cultivation.

Eventually all arable land was occupied. Land prices started moving up, as did prices of agricultural products. No longer could production be increased simply by acquiring fresh land. To enhance production and keep costs down, farmers started substituting horsepower and machines for the most expensive production factor, human labor. Another strategy was to farm with improved crop varieties and chemical additives.

Before the adoption of agriculture humans consumed thousands of species. One effect of the adoption of agriculture was the reduction of the number of plant species used as food. When people's welfare depends on one or a few crops they become highly vulnerable to crop failure. Ireland's great potato famine is the best example of the dangers of monoculture, but other examples abound.

The history of most Latin American countries is that of economies built on one or a few crops. For the last hundred years Argentina has relied mostly on wheat, maize, and linseed for its export earnings. Throughout its history Brazil has experienced boom and bust periods owing to its monocultures—sugar in colonial times, rubber and coffee since independence. Colombia is heavily dependent on coffee, Ecuador on bananas, Cuba on sugar, and so on. Elsewhere in the world Kenya relies on coffee, Sri Lanka on tea, Malaysia on rubber, Thailand on rice.

Monoculture has many negative social and economic consequences for the nations that practice it. A country that grows just one or a few crops becomes dangerously vulnerable to changes in market conditions. We have seen that in the case of Brazil, whose monoculture economies have ridden a rollercoaster in response to world markets since colonial times. Brazil should serve as a warning for countries to diversify the species they cultivate, for weather, disease, and natural or man-made catastrophes can lead to widespread death in a single-crop economy.

The agrarian revolution of the eighteenth and nineteenth centuries has received less attention than the industrial revolution, of which it was a forerunner. Yet the industrial revolution would not have been possible without a boost in agricultural productivity. Productivity rose because of an increase in farm specialization and the introduction of better technologies and farm machinery. Farmers turned to monocultures and the intensification of cultures, types of agriculture that are much harder on land resources than the extensive system of crop rotation that had prevailed until then. Standards of living began to improve, but the price was environmental deterioration.

Today we are witnessing a similar trend in tropical countries. To meet the demands of a growing population, farmers are enhancing productivity by adopting the approach of the European agrarian revolution. At present, this approach is displacing people from the land and swelling the already teeming slums of the third world.

Contemporary Farming

ONCE FARMERS ALL over the world were mostly self-sufficient. The popular image of a farm as portrayed in movies and children's books still conforms to that picture, with its blend of cows, pigs, horses, vegetable gardens, and fields. There are farmers who still grow most of the food they eat, but they generally live in remote places with poor communications, limited access to markets, and barren land. Most subsistence farmers today use traditional labor-intensive farming techniques. Because they are poor they cannot obtain credit and therefore suffer from a shortage of machinery, fertilizer, and other inputs that require cash.

Modern farmers, by contrast, are highly specialized and grow only a few crops for city markets. Specialization requires greater capital investment, more organization, more productive technologies, and machinery. Together these elements result in greater crop yields and increased productivity of land, labor, and capital. The modern farm is highly integrated with, and dependent on, city-based markets. Farmers have changed from self-sufficient individuals to agricultural industrialists.

A consequence of specialization and market-oriented production is the development of new industries. When farmers were largely self-sufficient there was no need for factories to make farm machinery. The local smithy made the hoes, shovels, and sickles. When farmers used animal manure there was no room for the manufacture of chemical fertilizer. When most produce was consumed on the farm, food processing was limited to basic processes such as milling, wine making, or olive pressing. These activities took place on or close to the farm. But when farmers started selling most of their produce to cities, whole industries sprang up in response to the need for elaborate farm products.

In turn, as city-based industries expanded, they offered steady employment for underemployed rural workers. As the proportion of population living and working in cities rose, the market for farm products grew, serving as an incentive for farmers to increase production and efficiency. Those farmers who adopted agricultural methods that reduced costs or raised production gained an advantage over other farmers, for as production increased the price of farm products decreased. This forced every farmer to adopt new technologies to stay competitive.

Today urban dwellers and the industries that depend on farm products have an interest in keeping their cost low.[1] Therefore they also favor highly productive agriculture. To increase productivity, governments encourage agricultural research and provide financial incentives. During the latter part of the last century in both Europe and the United States, governments set up agricultural experiment stations and established a system to deliver agricultural information to farmers. Universities began research programs and agronomy departments to increase and diffuse knowledge about the best farming methods. Today financial instruments such as crop insurance, farm credit, and futures markets reduce financial risks and ensure regular and steady production. Governments intervene by providing both direct and indirect subsidies. Subsidies are designed at times to encourage farm production, at other times to reduce planted acreage and decrease production. Sometimes governments want to guarantee minimum prices for farmers; sometimes they are interested in ensuring low prices for consumers. The result: Modern agriculture is very efficient and fulfills the objective of producing a steady supply of cheap food.

A curious consequence of all these changes is that while the absolute productivity of the economy's farm sector has increased, the farm sector's contribution to the total gross national product, its share of the labor force, and its growth in relation to the manufacturing sector have decreased. A comparison of the economies of countries at various levels of development illustrates this trend. Agricultural production accounted for 32 percent of GDP (gross domestic product) among the 39 countries with the lowest per-capita income in the world in 1985.[2] In the 56 countries classified as middle-income countries, agriculture accounted for only 15 percent of GDP, and in the 23 high-income countries it represented only 3 percent. Yet agricultural production per person was $422.81 in high-income countries, and only $205.77 and $79.74, respectively, in middle- and low-income countries. In high-income countries the average value of food produced per farmer is five times greater than in low-income countries, but agriculture represents only a thirtieth of the total production of high-income economies.

The change toward market-oriented agriculture has not come easily. Just as the enclosure movement in Europe increased productivity but created dislocations and social strains, so the change toward high-input, market-oriented agriculture in developing countries is creating social disruptions. This is exemplified by the "green revolution" in the Asiatic tropics.[3] This is only one of many dilemmas posed by the widespread adoption of modern farming. Perhaps the most challenging dilemma is this: How are we going to feed some additional five billion people in the next fifty years without degrading the environment on which agriculture depends?

Profit maximization was not the motivating force in traditional self-sufficient farming. The long-term objective of the farmer was the lowering of risk to ensure a steady supply of food for the family, for when crops failed the family was at risk. Crop diversification was the best way to reduce risk, and farmers favored crop varieties with broad tolerances that could grow well under various conditions. Often they cultivated more than one variety. Low-yield varieties were better than higher-yield but less tolerant ones. Even so, crop failures and famines were common in traditional farming communities.

Seeds came from the previous harvest; farm animals, crop rotation, and fallow restored soil fertility; and the farmer's family provided most of the labor. Farmers did not have to pay for labor, so they probably did not calculate labor costs or labor efficiency. Work was routine and physically exhausting. Farmers tried to find a middle road between ensuring enough production to satisfy the needs of the household and keeping the amount of drudgery to a minimum.

Typically, traditional farmers lived close to villages. Villages gave community support that brought with it constraints. Farmers near villages had access to community labor over short periods for raising a barn or harvesting a crop. Villages usually also had common grazing lands. In turn, village life determined crop types and farming methods. Institutions and people outside the village—for instance, landlords, the church, and the state—further restricted the freedom of the traditional farmer. They exacted part of the produce, either in kind or as money tribute or labor.

As we have seen, all of this began changing in Europe in the seventeenth century. The industrial revolution of the nineteenth century gave further impetus to agrarian restructuring. In other parts of the world, the type of change that took place in Europe in the eighteenth and nineteenth centuries has been taking place more recently. In the Americas, Spanish, Portuguese, and English colonization wiped out most native farming communities and laid the ground for market-oriented agriculture. Today North America, Australia, and New Zealand practice the most advanced farming in the world. Meanwhile, in Asia and especially Africa the two types of farming coexist: the traditional, village-centered, self-sufficient farms, and market-oriented farms producing mostly for export. But even there traditional farming has come under pressure from a spectrum of factors and is giving way to high-yield, high-input, market-oriented agriculture.

Traditionally, growth in food production has come from expansion of the cultivated surface. The process proceeded steadily for 10,000 years, then exploded over the last 150. According to historian J. F. Richards, 911.1 million hectares have gone under the plow in the last 130 years.[4] Only 59.6 million hectares reverted to noncropland in that period, for a net gain of 851.5 million hectares, or 57 percent of present arable land. With expansion no longer feasible, farmers have had to intensify cultivation to improve yields per unit surface. This involves the use of improved varieties, fertilizer, and year-round cultivation without fallow, collectively known as industrialized or "high-input" farming. Although it has taken root in much of the developed world, many questions about its sustainability are being raised.[5]

The agrarian revolution was not an offshoot of scientific knowledge. Responding to developing market forces, enterprising farmers first in Europe and later in the United States and other countries started experimenting.[6] They introduced new soil-improving crops such as clover and alfalfa, and they modified and eventually abandoned the traditional three-field rotation and bare fallows. Different kinds of organic and inorganic fertilizers were introduced, excessively wet soils were drained, new and better steel machinery was invented, and crop yields were increased by breeding better varieties. Only as interest in applying these new techniques grew did researchers begin to investigate the scientific bases of agriculture.

To produce a crop, farmers rely mostly on natural processes such as soil formation, rainfall, nutrient recycling, and control of pests by their natural enemies. Through observation and experimentation traditional farmers learned how to manipulate soil, water, and plants to ensure maturation of a crop. Knowledge was passed orally from generation to generation of farmers. More literate and intellectually inclined farmers, from Cato in antiquity to Thomas Jefferson in the Enlightenment, compiled agricultural knowledge and added their own observations.[7] By the middle of the nineteenth century botanists and zoologists began to investigate the behavior of crop plants, the relations between plants and their environments, and the principles of heredity, giving rise to the science of agronomy. The knowledge they acquired was applied to improve farming. Today agriculture can improve

natural growing conditions with better irrigation and chemical fertilization, control harmful insects and plant diseases with pesticides, and
increase yields with improved crop varieties. The new approaches
complement rather than replace natural processes.

How has greater understanding of the basic components of agriculture contributed to the structural transformation of agriculture?

Soil is one of the substances essential to crop growth. It is a complex
mixture of inorganic and organic materials, water, and air.[8] In addition, the soil is inhabited by many living organisms, from microscopic
bacteria to moles that constantly churn and modify the soil.

The inorganic particles derive from the slow decomposition of the
rocky mantle of the earth that underlies soil. Particles differ in mineral
composition depending on the rocks from which they come. The principal mineral component is usually silicate: According to size, particles
are classified as rocks and pebbles, sand (particles between 0.05 and 2
millimeters in diameter), silt (particles between 0.02 and 0.05 millimeter), and clay (below 0.02 millimeter). The characteristics and age
of the rock as well as the climate regime (mainly temperature and rainfall) are the principal factors that determine particle size. In dry climates such as deserts, coarse sandy soils are more prevalent, while in
wet climates such as tropical rainforests, fine-particle soils predominate.

The ratio of particles of different sizes gives soil its characteristic
texture. The best soil is loam, with an even mixture of sand, silt, and
clay, that is, a network of pores formed from the juxtaposition of particles of different sizes. When soil such as sand is predominantly com-
posed of big particles, the spaces between particles are too large to retain water well. Clay soil, on the other hand, tends to get waterlogged
and does not aerate roots and soil organisms sufficiently. Such soil,
called heavy, is hard to cultivate. Cultivation and machinery have the
effect of reducing pores and compacting the soil. Less intensive tillage
can reduce this problem.

The organic component of soil comes from decomposing plant and
animal remains, including leaf litter, roots, dead insects, and bacteria.
Organic matter provides many important nutrients, especially nitrogen. Soils poor in organic matter tend to be sterile.

The physical properties of soil can be modified to some extent by certain practices and inputs. Sandy soil can be improved by growing and then plowing under leguminous and other crops to increase organic matter. Such was the procedure to improve the "Norfolk sands" in England during the eighteenth century. Lime and marl can be added to reduce acidity in soil, and peat can be added to reduce pH in basic soil. Drainage tiles can be laid under low-lying areas where water accumulates to reduce waterlogging. Many soil improvement practices are too expensive to be justified in any but high-value crops. In the Brazilian savannas, temperature and rainfall are conducive to agriculture but the soil has a low nutrient content and high acidity. Treatment with lime to decrease acidity, followed by the addition of phosphate rock and nitrogen fertilizer, permits the cultivation of soybeans, wheat, and rice. The cost of soil additions are about $800 per hectare and must be repeated every ten to fifteen years (nitrogen fertilizer has to be added every year).[9]

Another important aspect of soil management is relief. One of the principal problems facing the modern farmer is soil erosion. Erosion is particularly serious in soils situated on a slope. Slopes can be terraced, an ancient practice, to reduce erosion, but terracing is expensive. Contour plowing, that is, making furrows parallel to the contours of a slope, can also reduce erosion significantly.

Soil management is an important aspect of modern farming, and land in many areas has been noticeably improved. The soil of Georgia and other regions of the southeastern United States, poor to start with and exhausted from cotton and tobacco monocultures in the last century, has been restored through such techniques as planting leguminous crops and applying inorganic fertilizer. Nevertheless, many soil problems remain, primary among them the loss of topsoil to wind and water. Calculations put the loss of arable soil at 5 to 7 million hectares per year.[10] Erosion results from overuse of soil and other inappropriate practices in heavily populated areas, and from cultivation of marginal lands, for example, in dry regions such as sub-Saharan Africa. In many tropical regions, an alarming loss of soil has resulted from the spread of small-scale peasant farming into mountainous or arid areas not suited to their traditional techniques. On the other hand, the careful terracing and management of mountain slopes has resulted in productive, sustainable agriculture in many regions of Asia.

Water is as essential as soil to the growth of plants. Through their leaves they transpire 99 percent of the water their roots take up, incorporating 1 percent. If the soil around the roots gets too dry a plant wilts, and if the condition persists it dies.

Most agriculture depends on rainfall to supply water to plants. Since the early days of agriculture, people in dry areas or areas with unpredictable rainfall have taken water from rivers, lakes, and wells to supplement rainfall. Irrigation projects are among the major contributors to the growth of agriculture in the last half century. Yet in many parts of the world there is little spare water left, and in areas such as California urban centers and farms must compete for dwindling supplies. Conservation is the only answer. New techniques, for instance, drip irrigation and the use of plastic sheeting to reduce soil evaporation, are increasing the efficiency of water use. Yet in areas of shortage the cost of water may rise so high that farmers have to cut back planting.

In the past, some farmers had to abandon agriculture because of the salinization of the soil, a result of improper irrigation. All water contains some salts in solution. As water evaporates it leaves the salts behind. They accumulate especially in the upper layers of the soil and impair plant growth. Eventually crop yields are so low as to make farming futile. This first happened in Mesopotamia 3,000 years ago, and today it is happening all over, for example, in California, Mexico, and India.

The major plant nutrients are carbon, oxygen, hydrogen, nitrogen, phosphorus, potassium, sulfur, and sodium. Plants obtain carbon from air, hydrogen and oxygen from air and water. All their other nutrients—representing almost every mineral element on earth—come from the soil.

In natural ecosystems, most soil nutrients cycle from the soil to the plants, to the microorganisms, fungi, and animals that eat the plants, and then back to the soil. Nitrogen and carbon have slightly more complex cycles. New nutrients are added through the decomposition of parent rock, debris brought by wind and water, and nutrients dissolved in rainwater. Farming, especially modern, removes the nutrients in crops from the ecosystem and transports them to cities,

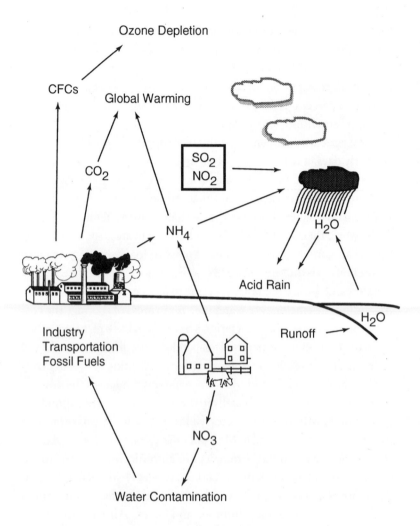

Ozone Depletion

CFCs

Global Warming

CO_2

SO_2
NO_2

NH_4

H_2O

Acid Rain

H_2O

Industry
Transportation
Fossil Fuels

Runoff

NO_3

Water Contamination

Farming and its associated activities are significant sources of greenhouse gases [methane (NH_4) and carbon dioxide (CO_2)], water-polluting nitrates (NO_3), oxides of sulfur (SO_2) and nitrogen (NO_2) that are major contributors to acid rain, and chlorofluorocarbons (CFCs) that contribute to the depletion of the ozone layer. Thus, farm output is hurt by the effects of these environmental changes that farming itself helps to create. Acid rain hurts crop growth, increased ultraviolet radiation due to ozone depletion harms crops and farm animals, and global warming has the potential to fundamentally alter the environment and threaten global crop growth patterns.

where they eventually end up in sewage. Thus to keep soils fertile, nutrients have to be replenished through fertilization.

Animal manure has been used since the early days of agriculture to improve soil fertility. Leguminous crops have been grown toward the same end. The litter and roots of these crops, when incorporated into soil, make the nitrogen they have taken from the air available to other plants. Although legumes increase the nitrogen level in the soil, they consume phosphorus and potassium, which must be supplied to the soil.

As crop yields increased during the eighteenth and nineteenth centuries, more nutrients were removed from the soil and exported to the city. A number of materials were brought in to supplement the farm's animal manure—the contents of privies and cesspools, sweepings from stables and streets in nearby cities, and so on. By the 1830s water transport had become cheap enough that Peruvian guano—bird droppings from offshore islands—could be imported into Europe for use as fertilizer. This was supplemented by inorganic nitrates from Chile in the second half of the nineteenth century.[11] Eventually scientists learned an industrial method of taking nitrogen from the air.

By the end of World War II, cheap nitrogenous fertilizer could be industrially produced. For the first time in history, fixed nitrogen from the air became both abundant and relatively cheap. Now it was possible to apply nitrogenous fertilizer to enormous areas. The cereals were among the most responsive crops (see table 2). Yields have continued to increase, primarily in industrial nations capable of widespread fertilizer production but lately also in developing nations that import nitrogenous fertilizer or produce it themselves.

Although nitrogen usually represents the principal deficiency in soil, phosphorus and potassium are also in short supply. Phosphorus was traditionally provided through manure; crushed bones are also a good source. In the 1840s, "superphosphates" were introduced in England, the result of treating phosphate rock with sulfuric acid. This was the first artificial fertilizer and is still the major source of phosphorous fertilizer.

Inorganic fertilizer in large measure accounts for the increase in agricultural production since World War II (table 3). Modern high-yield

Table 2

Maize and Wheat Yields and Fertilizer Use by Region, 1980–82
(listed in ascending order of fertilizer use, in kg/ha)

Region	Maize Yields	Wheat Yields	Average Fertilizer
Africa	13.3	11.3	19.5
South America	20.1	14.1	26.9
Australia	—	10.3	27.9
Asia	23.4	17.3	68.5
U.S.S.R.	28.1	14.9	82.6
North and Central America	54.7	22.9	88.4
Europe	47.8	37.7	225.0
World average	33.2	19.3	78.5

Source: FAO *Production Yearbooks*

crops have been bred to respond positively to additions of inorganic fertilizer. Only a fraction of the inorganic fertilizer applied to a field is absorbed by the crop or retained in the soil. The remainder is dissolved in rainwater and washed away into creeks, rivers, lakes, and oceans. Some can also find its way into groundwater. Increased nutrients encourage the growth of aquatic plants and algae that can rob water of its oxygen and result in the death of fish and other aquatic life, a process known as eutrophication. A high nutrient content in drinking water, especially a high nitrate concentration, is harmful to human health. This is a serious problem in areas of intense agriculture such as the Netherlands and parts of the United States.

An indirect effect of inorganic fertilizer has been to increase the disparity between rich and poor countries. Rich countries have the industrial base to produce reasonably priced fertilizer, which is not always the case in poor countries. Consequently farmers in rich countries can more easily increase their productivity. Without inorganic fertilizer, the United States could not be the big producer and exporter of grain that it is. Without it, Europe would be a big importer of food.

Table 3
World Consumption of Inorganic Fertilizer Nutrients
(in millions of tons)

Year	N	P$_2$O$_5$	K$_2$O	Total
1938–39	2.6	3.6	2.8	9.0
1953–54	5.2	6.3	5.7	17.2
1959–60	9.7	9.7	8.6	28.0
1965	16.6	14.0	10.9	41.5
1967	22.1	16.3	12.9	51.3
1969	27.1	18.3	14.5	59.9
1971	31.8	19.8	16.4	68.0
1973	35.8	22.5	18.5	76.8
1975	38.8	23.0	19.5	81.3
1977	46.1	26.6	22.9	95.6
1978–79	53.7	29.9	23.9	107.5
1980–81	60.7	32.8	25.2	118.7
1981–82	60.4	30.9	23.9	115.3

Source: FAO *Fertilizer Yearbooks*

Breeding, like fertilizer, has contributed to the structural transformation of farming. Knowledgeable farmers always selected the strongest, healthiest, and most productive plants in a field as the source of seeds for the following year's crop. However, only with the unraveling of genetic principles at the beginning of the twentieth century did scientists come to understand the basis of plant selection, making possible planned breeding of high-yield varieties with desirable characteristics. Agricultural colleges and research centers supported efforts to breed new varieties. New crop varieties and the better understanding of plant nutrition, which led to improved soil management and the application of chemical fertilizer, significantly increased agricultural productivity. Wheat and then maize yields rose rapidly in Europe, the United States, Canada, Australia, and Argentina.

Traditional farmers reserved a part of the harvest for future plantings. Today most farmers buy certified seed from specialized companies. This practice is mandatory if they want high-yield hybrids. Certified seeds constitute a major expense and a real break with the age-old tradition of producing seed on the farm. The modern farm is dependent on outside sources for its most important input. [12]

Although plant breeding has brought remarkable results, it has undesirable effects. A major problem is the loss of genetic diversity, or "genetic erosion." Modern high-yield varieties also have more demanding requirements in the way of water, temperature, and nutrients, and often they are more susceptible to weeds, pests, and disease. Their cultivation is more exacting and requires large expenditures in fertilizer, herbicides, and pesticides.

Crops are attacked by many animals that eat plant material, from microscopic nematodes to birds, rabbits, and deer. But the major enemies of crops are plant-eating insects, especially caterpillars, bugs, and beetles. Crops are also susceptible to diseases—most of which are produced not by bacteria but rather by fungi or viruses. Weeds, which compete for nutrients, water, and sunlight, reduce crop productivity.

Pests and weeds were not a serious problem in traditional agriculture. Hoeing and harrowing removed some weeds, but traditional land races were better equipped to compete against weeds than modern high-yield varieties. Rotation of crops and fallows as well as natural enemies kept pest populations and diseases from accumulating. Today's practice of planting large areas with a single crop has created an environment that favors the growth of insect populations. The higher nitrogen content in fertilized crop plants also favors pests. Pest outbreaks can totally destroy a crop, making artificial control essential.

Many wild plants produce compounds that protect them from pests. Because these compounds are often eliminated in the process of domestication, crops are less resistant than their wild ancestors. To protect crops, agronomists introduced insecticides, compounds that poison insects. The first insecticides were plant products such as pyrethrum, which comes from the plant of the same name, and nicotine, from the tobacco plant. Over the last fifty years scientists have developed many synthetic chemicals to combat insects as well as weeds and plant disease.

Natural organic compounds have a short life, quickly breaking down into harmless substances. To be effective they have to be applied often, which drives up the cost of protection. Synthetic stable compounds like DDT were introduced after World War II. Unfortunately, they proved to have an even greater drawback: They killed not only pests but also beneficial insects such as those that pollinate crops. Moreover, they killed animals further up the food chain, including birds and other vertebrates that eat insects, and in many cases were thought to endanger human health. The reduction in natural enemies removed an effective natural control of crop pests. On top of all that, many pests began to develop resistance to synthetic compounds, while predator insects did not. The loss of natural controls meant that even more insecticide had to be applied, increasing costs and endangering the lives of countless more harmless creatures.

The costs of insect attack and chemical control are high. In the United States insects create losses of $0.5 billion annually in cotton alone, and more than $150 million is spent annually to control them.[13] Additional losses to pests take place during transportation and storage, to the tune of more than 30 percent of all agricultural production in the United States.[14] Meanwhile, cotton pests have become resistant to some insecticides, and there are serious environmental and health problems associated with the use of cotton insecticides. After cotton, maize receives the largest dose of chemicals to control insects and weeds. Apples receive more pesticide than any other fruit crop, and on a per-acre basis more than any other crop. However, losses due to insects have not been reduced for these crops.

Modern agriculture is faced with a dilemma. Unless they are protected, crops are vulnerable to pests and disease. But chemical pesticides do not solve the problem. They are damaging, expensive, and not always cost-effective. Experts feel that the time has come to stop relying on chemicals and turn to more environmentally friendly approaches to insect control.

Sophisticated machinery was yet another factor in the remaking of agriculture. At the beginning of the eighteenth century, farmers employed the same tools as medieval farmers: wooden plows, hand hoes, and harrows for cultivating, sickles, scythes, and cradles for harvesting, and flails for threshing. Agriculture was a backbreaking enterprise the extent of which was determined by the size of the family, except on plantations employing slave labor.

As America's Midwest and West were opened to agriculture, farmers encountered soil too heavy for the wooden plow. In 1837, building on others' experiments with cast-iron and steel plows, John Deere of Illinois built a one-piece wrought-iron plow with a share made of steel. The Deere plow was so efficient in turning the heavy Illinois soil that it got nicknamed the singing plow. By the mid-1850s 10,000 of them were being produced annually. Today wooden plows survive only in traditional agriculture, especially in stony soil.

The most labor-intensive operation of the farm year is the harvest, which must proceed rapidly once the crop is ripe. Until harvesting could be made more efficient, no real advance in labor productivity was possible. After many independent attempts in the United States and Europe to produce a reliable and efficient mechanical reaper, Cyrus McCormick developed one in the 1850s. It soon became popular, making the McCormick Company the leading manufacturer of reapers.

Almost simultaneously, threshing machines were invented in Europe and the United States. At the beginning they only threshed; straw, used for various purposes, including animal feed, still had to be separated by hand. The grain likewise had to be independently winnowed to separate the grain from the chaff. Eventually the J. I. Case and Pitt companies in the United States developed equipment that combined all these operations. The machines were large, cumbersome, and expensive. Because ordinary farmers could not afford them, specialized itinerant operators would charge a fee to thresh and winnow grain with the new machines. Twenty years later machines called combines appeared. Pulled by as many as forty horses, these cut, threshed, and winnowed. Today all grain harvesting is done with self-propelled combines.

Once the harvest had been mechanized, horse-drawn cultivation machines like harrows, planters, and mowers appeared on the market. Each of these machines represented a considerable investment. Labor shortages speeded up the adoption of farming machinery, rapidly accelerating per-person output.

The last stage of mechanization was the replacement of horses by tractors. Tractors appeared in the last decade of the nineteenth century and by 1910 were firmly established. Trucks and self-propelled har-

A horse-drawn combine in California. It took teams of sixteen horses to pull such a heavy machine.

vesting equipment followed. Finally, cheap fuel enabled the internal combustion engine to be used in agricultural machines. Increasingly, these changes made agriculture more productive as well as capital intensive, as one California farmer recently made clear: "A retired neighbor of mine was doing some planting for me with our new 13-foot, 6-inch drill, which takes 1800 pounds to fill the seed hoppers. He started at 6:30 and went home at 5:30, and planted over 75 acres of grain. He was just shaking his head, because he said that if he planted 20 acres a day in the early 50s, he'd had a big day. So now it's easier, but it's capital intensive."[15]

The program of introducing high-input agricultural techniques to certain parts of the third world to increase yields is called the green revolution. This transfer to the tropics of temperate-zone agriculture illustrates both the accomplishments and the problems of modern agriculture.

The areas most affected by the green revolution are Asia, Africa, and certain parts of Latin America. The revolution has its origin in a pilot program of the Rockefeller Foundation to produce higher-yield maize and wheat varieties. Initial research was done at the agricultural center of the University of Mexico in Chapingo. Its success led the Mexican government and the foundation to create the International Maize and Wheat Improvement Center in 1966. A rice center was later set up at the University of the Philippines at Los Baños, where the Ford and Rockefeller foundations, together with the government of the Philippines, created the International Rice Research Institute in Los Baños.

In 1971 the Food and Agriculture Organization (FAO), the World Bank, and the United Nations Development Program (UNDP) sponsored the formation of a network of agricultural experiment stations around the world. A consortium, the International Consultative Group for Agronomic Research (ICGAR), administers the network. Among the members of ICGAR figure three regional development banks, the European Fund for Economic Development, the governments of thirteen nations, and three private foundations. Besides the two mentioned, research centers include the International Potato Center in Peru, started in 1974; the International Institute of Tropical Agriculture in Ibadan, Nigeria, established in 1967; the International Center for Tropical Agriculture in Cali, Colombia, also established in 1967; the International Crops Research Institute for the Semi-Arid Tropics at Hyderabad, India, established in 1972; and in Ethiopia the Center for Research on Cattle in Africa.

Much of the third world is located in the tropics. High-input, high-yield agriculture was developed for the physical and climatic conditions of the temperate zone. Its transfer without modification to the tropics is not possible on account of differences in climate and soil. In the tropics, heavy rainfall combined with old geological formations has produced soil generally poorer in nutrients than temperate soil, and it has increased erosion in exposed soil. Moreover, mild temperatures year-round favor weed and pest populations.

People living in the tropics have perfected swidden agriculture to lessen these problems. Slash and burn is ecologically sound but requires much more land per person than temperate agriculture for similar yields. Tropical populations have also perfected high-yield systems in areas with excess water, paddy rice in Asia being one example. Especially after World War II it became apparent that traditional agriculture would not suffice to feed the burgeoning populations of the third world. A food shortage would develop unless agriculture became more productive. This was the issue addressed by ICGAR and the scientists at experiment stations.

One possibility would have been to start with tropical methods and increase their productivity. The strategy chosen instead was to apply the methods of temperate agriculture in the tropics. High-yield varieties adapted to tropical climates were developed and grown with generous doses of fertilizer. Chemical herbicides and pesticides controlled weeds and pests. This technological package has been uniquely successful in improving productivity, which all over the third world, especially in the tropics, has grown faster than population. This has solved, at least for the moment, the general food problem. But the green revolution has not been without problems or critics.

The new high-yield varieties, also called miracle varieties, require near-optimal conditions for success. That is, they need adequate water, good soil, generous fertilizer, and protection from weeds and insect pests—all costly. Their successful cultivation also requires agronomic knowledge. Thus only the better placed and educated farmers can take full advantage of the benefits of high-yield crops. To cover costs, farmers have to sell their product for a good price in the market. They concentrate on growing the most profitable cash crops. The increased profits accruing to farmers who adopt new varieties allow them to buy more land and increase their holdings. Farmers on more marginal land, without adequate schooling or the ability to invest in seed, fertilizer, and pesticide, see their income reduced when prices drop as a result of the success of richer farmers. Many sell their land and become salaried workers on more prosperous farms; others emigrate to cities, where they often join the ranks of the poor. Thus, while the green revolution has increased farm production—countries such as

India have seen their need for food imports drastically reduced and have sometimes turned from importers into exporters—it has also resulted in social inequality. Poor farmers have lost the safety net that existed in mostly subsistence farming villages before the green revolution. Successful farmers increasingly see their land as a business rather than a way of life. The changes occurring are in many ways reminiscent of the shifts brought about by the agrarian revolution in seventeenth- and eighteenth-century England.

It is impossible to say whether these changes will be beneficial in the long term. Success depends on halting population growth and finding employment for the city poor, as in nineteenth-century Europe. Some countries appear to be succeeding; many others are less capable of absorbing the displaced peasant population.

The green revolution is also potentially harmful to the environment. The increased use of pesticides and fertilizers threatens aquifers and water supplies. Many insect pests are acquiring resistance to pesticides, which could spell disaster for this agricultural strategy. [16] The commercialization of agriculture in the tropics favors crops that command a higher price, such as sugar, coffee, and tea. Less surface is dedicated to growing cheap food plants such as millet or manioc, which increases the cost of food for the poorest members of the society.

As scientists survey the history of food procurement over the last 10,000 years, they notice an accelerating increase in the rate of landscape transformation. Gone are most of the earth's virgin forest, most of the natural grassland, much of the shrub and swampland, and large parts of the natural savanna. Where forests still stand they have been profoundly affected by humans. Not all landscapes are affected equally. The Middle East, Europe, India, and the Far East, especially China and Japan, are much more modified than most tropical areas. But the latter are now being quickly transformed.

For almost a million years little changed in the manner in which hunter-gatherer people obtained their food. [17] With the beginning of agriculture came the first stages of human-made environmental transformation. At first the changes were not significant. It took almost 5,000 years after the invention of agriculture for cities to arise and change the relation of farmers to the rest of society. Then farming expanded and landscape transformation accelerated. For the next 3,000 to 4,000 years this relation between city and farm was maintained.

Four hundred years ago the relation of farmers to cities changed again. Slowly farming began to concentrate on producing an excess for consumption by people living in cities. With the industrial and agrarian revolutions in the last century and a half, the emphasis turned to the production of surpluses for a commodity market. The demands of the industrial revolution, and the new machines, accelerated the rate of environmental transformation. These developments occurred first in Europe, then spread to the rest of the world. Today, farming depends throughout much of the world on machinery, fuel, and chemicals, means of cultivation that severely disturb the natural environment.

In the 8,000 years that have elapsed since agriculture was introduced into Europe, landscapes have been drastically transformed. In the United States the same process has taken place in a shorter time. When Europeans first came, forest covered one half of the land. More than half the forests east of the Mississippi and a quarter of those to the west have been cleared to make room for agriculture. Most of the tall-grass prairie that once covered the Midwest has been transformed into arable fields. That land is now among the best and most productive cropland in the world. The best areas in the short-grass prairie farther west are now used for either irrigated or dryland farming. Much of the land in the dry West is used for grazing. Similar transformations have taken place in Russia, Argentina, Australia, and Japan.

The high-yield agriculture that reigns throughout much of the world and depends on inputs of energy, capital, fertilizers, and pesticide transforms more than the land; it transforms aquifers and the atmosphere. Is this type of cultivation, which in its short existence has had such long-term effects, sustainable?

A sustainable process is one that can be maintained at a certain optimal level for a long time, theoretically forever. The question is whether the present system of high-input agriculture is optimal, or whether it will have negative side effects that eventually reduce its output. This is a critical matter: If modern agriculture is not sustainable, humanity could be faced with a food crisis in the future.

Agriculture has always relied on some inputs from outside the farm, such as nails, salt, and iron implements. With the rise of industrialized society and commercial farming, inputs have increased dramatically to include fertilizer, pesticide, machinery, and hired labor. In the United States, outside inputs have gone from 45 percent of gross farm income in 1900 to over 80 percent today.[18] Between 1950 and 1985, the share of total farm costs represented by manufactured inputs, interest on loans, and capital expenses rose from 22 to 42 percent, while the share of labor and inputs of farm origin declined from 52 to 34 percent. The same is happening in other parts of the world, particularly in those areas of the third world that today show the greatest gains in productivity. This indicates that farms are as dependent on cities, from which most inputs come, as cities are on farms. Can such interdependency be maintained without incurring environmental or other effects that will reduce output? The main concerns are groundwater contamination, food quality, farm workers' health and safety, the effect of farming on natural flora and fauna, and the dwindling supply of oil.

The dramatic rise in the use of chemical products in farming could be seriously harmful. Many of these substances are toxic and degrade slowly. They enter the food chain either directly through chemical residues in food or by contaminating groundwater used for drinking. Furthermore, they pose a health hazard to farm workers who have to handle them. A number of well-substantiated cases exist of chemicals being present at dangerous levels in food and water, as well as cases of farm workers being poisoned.[19] Chemicals have contaminated natural environments and endangered wildlife. There is also concern that the oil reserves of the world are dwindling, and that eventually oil will become so expensive as to preclude its use in agriculture. All of these problems together suggest that present-day agricultural methodology is not optimal.

Other worrisome questions raised in relation to the sustainability of modern agriculture relate to soil erosion, and the loss of genetic variability in crops. Agricultural scientists are trying to devise economical means to reduce the dependency of agriculture on some of these risky inputs. Biological control of pests, low tillage, organic farming, improved timing of planting, and resistant varieties are some of the promising methods being explored to reduce the heavy dependence of agriculture on chemical inputs and to improve the natural landscape. Time will tell whether the system can be made truly sustainable.

During the 10,000 years that humans have farmed, a close relationship has existed between population growth and agricultural output. We do not know whether population growth is checked by the amount of food, as Thomas Malthus thought, or whether it is the motivating force for agricultural growth, as Esther Boserup and others have suggested.[20] Researchers are inclined to see the two factors as interacting in a feedback system. Agriculture most assuredly arose in the Middle East as a response to a need for increased food supplies in a growing population unable to find ways of widening its foraging territory. In classical and medieval times cultivation slowly replaced hunting and gathering because it generated higher yields per unit surface. As the European population slowly increased, agricultural output expanded through extension of the arable surface until the limit of existing technology was reached. The crisis of the fourteenth century might be an example of Malthusian population control. Society responded with a reorganization of the system of agricultural production and social organization, and soon after that both food production and population resumed their upward movement.

Humans lived as hunter-gatherers for over two million years. Though around for only one percent of that time, agriculturalists have managed to thoroughly transform the world's landscape and, according to many, endanger the existence of life on this planet. The challenge now is to find a way of stabilizing both the population and the food supply to make agriculture as sustainable as the hunting-gathering way of life once was.

C H A P T E R

The Future of Food

FOR THE LAST 10,000 years humankind has steadily increased the amount of earth's surface under its direct or indirect control. Fields, forests, meadows, lakes, rivers, mountains, coastal waters, deep oceans—soon there will hardly be a place that is not modified and managed by people. With control there comes responsibility, but people have not been able or skillful stewards of nature. They have watched species of plants and animals go extinct, permitted millions of tons of topsoil to erode, destroyed forests, and dumped billions of tons of chemicals into the air and water. Past disregard of the effects of environmental modification is today threatening humankind's future.

According to the UN population division, at some time in the coming century there will be twice as many people on earth as there are to-day, requiring at least twice as much food, clothing, and housing.[1] Much of the earth's population is not adequately fed, clothed, or housed today, and there is no assurance that the basic means of survival can be provided for an increased human population.

The task before us is daunting. We must reverse the process of landscape degradation while simultaneously doubling the amount that we extract from the landscape. If we cannot meet these goals, there is likely to be strife, suffering, and misery as people fight not to starve. Since population growth will be concentrated in tropical countries, many of which are not blessed with conditions conducive to agriculture, food shortages might also spark war between countries. Can this formidable challenge be met, and in the face of human-made climate change?

New and better farming methods will solve some of the problems in the short run. Farmers will have to pay special attention to the way agricultural methods affect land, water, and genetic resources.[2] They will have to combat pests and diseases in ways that are not harmful to people and animals. Biotechnology applied to agriculture may also solve some problems. However, in the long run two further changes in human society are needed: stabilization of human population and a more equitable distribution of income. Economist Michael Young of Australia has referred to these solutions collectively as the three Es: economic efficiency, environmental integrity, and equity.[3]

Efficient allocation and use of resources tend to increase total production. We are accustomed to thinking about efficiency at the level of the individual enterprise, where labor and capital are factors that directly affect productivity. But policymakers must also concern themselves with the economic efficiency of society as a whole. In making economically efficient decisions about land use, it is important for policymakers to include or at least estimate the value of unpriced benefits such as wildlife amenity and landscape diversity. An important group of unpriced benefits might be called ecosystem services, that is, contributions that a given piece of land makes toward the maintenance of the biosphere. Consider marshes. For a long time people have seen them as nuisances, barriers to communication and growth and sources of disease. But they are also home to a wide variety of plants and animals that would become extinct without them. They also play a role in flood control. To determine the best use of marshes or any other area, we must assign such ecological functions an appropriate value.

The ecological time scale of economic activity is also important. The aim should be not to mine land efficiently but rather to maintain and where possible enhance its long-term productivity while preserving the ecological services it provides to the biosphere. Economic efficiency is as much about the efficient allocation of resources as it is about use. The world is littered with examples of land-tenure systems that, because resource rights are not allocated efficiently, are being used ineffectively.

Unless environmental integrity is maintained, future generations will be compromised. Unlike economic efficiency, environmental integrity focuses on absolute as opposed to relative scarcity. It addresses the question of the optimal scope of human activity that allows ecological systems to continue functioning. The aim is to ensure natural resources to future generations. Any activity that eliminates a resource is suspect. Some economists assume that substitutes for most environmental functions exist. No evidence exists to corroborate this assumption. For example, no substitute for the tropospheric ozone layer is known. Given this and the concern for coming generations, it is essential that the integrity of existing ecosystems be maintained.

The desire for social equity brings in a different policy perspective. Most governments are uncomfortable with policies that lock the poor in place while the rich get richer. However, present policies appear to have just this effect. Ways will have to be found to distribute food, clothing, and shelter more equitably among people and nations if massive disruption is to be averted. Existing national and international institutions, especially agricultural ones, will have to be empowered in their capacity to resolve problems.

People consume different amounts of food of differing quality depending on their level of income. Poor people not only eat less, they eat lower-quality food such as tubers and cereals. Rich people consume more meat, up to a third of their diet. If general living standards improve during the next hundred years, the public will want not only more but better-quality food. For farmers to supply the projected food needed by future generations, they will have to produce on average 2 percent more each year during the next century and a half. To increase food production this much, and to do it sustainably, is the greatest challenge that farmers have ever faced. For the last thirty years average increases have been close to 3 percent. However, when farmers use up the best land and implement modern farming methods worldwide the growth rate of agricultural production will decline unless agronomists devise more appropriate methods. Soil conservation, minimum tillage, drip irrigation, improved high-yield varieties, integrated pest management, and biotechnology are some of the new techniques that may help maintain sustainable growth in agriculture.

During the last fifty years, agricultural production has kept up with unprecedented growth in the human need for food and fiber because farmers enlarged cultivated areas as well as increased crop yields. Unless crop yields continue to increase along with population, the world faces a serious food crisis. It is not known whether farmers can increase yields much with present technology, and many experts doubt that it can be done sustainably.[4]

To increase yields, agronomists must address a number of issues, among them the adverse effects of farming practices on sustainability, such as loss of topsoil from land bared by plowing; salinization and loss of fertility due to inadequate irrigation practices; the distortion of markets caused by subsidies and tariffs; the impact of global climatic change on agricultural yields; the contamination of water by pesticide and fertilizer; damage to wildlife from unsafe methods of combatting agricultural pests; and the loss of biodiversity.

There are approximately 11.5 billion hectares of land that people today can use for agriculture, livestock raising, forestry, and industry. Of these, 500 million hectares present no physical impediment for growing crops. Another 2.7 billion hectares require conditioning—fertilization, terracing, or irrigation—before cultivation. The remaining area, approximately 8.2 billion hectares, is suited only for forestry or livestock raising because it is too dry, too wet, too steep, or too infer-

Table 4
Land Use (in billions of hectares)

Category	1882	1991	Percent Change
Arable land and permanent crops	0.86	1.5	+74
Permanent pastures	1.5	3.2	+113
Forest and woodland	5.2	4.1	−21
Total	7.56	8.8	+16

Source: D. C. Norse *et al.*, 1992

tile. Farmers presently raise crops on 1.25 billion hectares, about half of the area suitable for raising crops (see table 4). This includes all the best soil—that which needs the least investment to make it suitable—as well as some marginal land being used inappropriately and in imminent danger of degradation.[5]

Soil improperly tended degrades. When wind blows or water washes away fertile upper soil, what is left can no longer hold sufficient moisture or support the same amount of plant growth. Soil also deteriorates when salts accumulate in irrigated lands or when heavy machinery compacts the earth. These phenomena are respectively referred to as desertification or aridification, salinization, and compaction.

Another danger is the loss of good agricultural land to cities, roads, and factories. Although the areas cities gobble up are not large in absolute terms, the impact of expansion can be considerable because prime agricultural land surrounds most large urban centers.[6] The problem is acute in countries where there is a shortage of agricultural land, such as the Netherlands or Japan.

Clearing tropical forestland for agriculture has created another set of problems. Many forest soils in the tropics are poorly suited for continuous agriculture, because they are not fertile enough or the terrain is too steep. Use of these soils requires careful assessment of their potential and special measures such as terracing or intercropping to avoid erosion and loss of fertility. Deforestation in the upper reaches of watersheds results in periods of flooding during the rainy season, followed by drought in the absence of a forest buffer. This creates serious difficulties for agriculture in floodplains and valley bottomlands, as either too much or too little water reaches these areas during the growing season. For example, in India more than 20 million hectares of agricultural land are flooded annually as a result of deforestation in the Himalayan region. In the Ganges plain alone, flood damage regularly exceeds $1 billion a year.[7]

Soil erosion is a difficult process to evaluate. Some erosion always takes place, especially in upland areas. The great agricultural districts of the world, such as the Ukraine, the U.S. corn belt, and the Argentine pampas, were all formed by sediments deposited from surrounding high areas. Soil, which is formed and removed constantly, is part of a dynamic system. The rate of erosion increases when vegetation is removed for purposes of agriculture. A serious problem arises when the rate of removal exceeds the rate of formation. Unfortunately, when damage to the soil starts affecting plant yield, it is often too late to do anything.[8]

There is no precise worldwide data on the actual loss of topsoil to erosion. The FAO estimates that 5 to 7 million hectares, approximately 0.3 to 0.5 percent of the earth's total arable land, are lost each year.[9] According to UNEP, close to 2 billion hectares of land are threatened by aridification due to loss of soil humidity. Much of this land is rangeland rather than cropland. The problem is general and there are few regions free of it. The seriousness of erosion differs from region to region. The problem is most pressing in the developing world, where the pressure on land is greatest. But it is also serious in the developed world, including areas with the richest soils. Some estimate that in the developing world unchecked erosion could reduce the area of potential cropland by 18 percent over the next ten years.[10] Crop productivity could be reduced by 29 percent. Experts estimate that the indirect costs of land degradation in Ghana are 7 percent of GNP and in Nigeria 17 percent.[11] Studies done in 1980 suggest that, if the rate of erosion in the United States in 1977 were to continue for fifty years, crop yields would be 8 percent lower than if there were no erosion.[12] Soil conservation should become the first priority of agriculture worldwide if yields are to increase and agriculture is to be sustainable.

The basic principles for controlling soil erosion are simple. Farmers should always keep their land covered with vegetation. Cultivation should take place only on flat land. And landowners should build barriers to impede the free flow of water and wind. Implementation of these principles is not so simple.

Control of soil erosion requires change in farming practices and investment in works. Investment can be sizable. For example, terracing in mountainous areas, essential to reducing soil loss, is expensive and requires an infrastructure of roads. But infrastructure itself can cause erosion, and rebuilding or maintaining it is often beyond the budget of poor countries.

The problem of soil erosion has given birth to a variety of new cropping practices collectively known as conservation tillage. The basic principle of conservation tillage is to leave enough crop residue on the soil surface to reduce erosion. Residue is the vegetable matter left in the field after the crop is harvested. It consists mostly of green or dry stems, leaves, and weeds. The amount and quality of residue vary with the crop and the method of harvest.

The idea of leaving crop residue in place is not new. Domestic animals, primarily cattle and sheep, can graze on the stubble, as is customary in traditional but rare in industrial agriculture. This procedure keeps a cover on the ground, helps feed the animals, and adds some manure to the field. The principal drawback is that animals, especially cattle, compact the soil surface. Another alternative is to leave stubble in the field and allow weeds and native plants to grow until the farmer is ready to plow. Dead plant matter decomposes and increases the organic matter in the soil. This process can be hastened by chopping up residue and adding chemicals to hasten the process of decomposition. Another alternative is to burn the residue and incorporate the ashes into the soil. This procedure, which destroys many of the nutrients and pollutes the air, is not recommended. The most effective method of all is conservation tillage.

Conservation tillage involves three basic approaches: "no-till," "ridge-till," and "mulch-till." In no-till agriculture, seeds are planted directly in the stubble of the old crop with special drills that make small holes in the soil for sowing. In ridge-till, seeds are sown on ridges created in the previous growing season and left undisturbed since the harvest. In mulch-till, farmers plow or disk the growing surface as in conventional agriculture, then cover the bare soil with some of the residue from the previous crop (mulch). Weeds are controlled with herbicides. Conservation tillage has the advantages of reducing labor and fuel and promoting wildlife.[13] It also increases yields. In the United States, wheat yields have gone from 1,167 kilograms per hectare from 1947 to 1975, when conventional tillage methods were used, to more than 2,200 kilograms per hectare in the period after 1975, when conservation tillage became widespread. (Not all the yield increase can be attributed to the change in tillage methods.)

Soil erosion cannot be controlled without active participation of the public sector. Government action can take many forms. Examples are direct participation in the building of erosion-control works, subsidies and tax incentives to motivate farmers to invest in erosion-control measures, programs to educate farmers on the problems of erosion and how to control it, and legislation to encourage appropriate land use.

Agricultural production averages twice as much on irrigated lands as on unirrigated lands. However, the introduction of water into an arid zone can have unforeseen consequences for both plants and soil. As we have seen, all water contains some dissolved salt, the amount varying from less than 100 milligrams per liter to more than 1,500. The less salt water contains, the more desirable it is for irrigation. Irrigation water comes from permanent rivers and underground aquifers.

Irrigation is practiced when rainfall is either too scarce to raise a crop or not sufficiently predictable during the growing season. However, importing water inevitably raises the watertable and the level of salts in the soil. The problems are serious. It has been estimated that in India alone more than 20 million hectares have been damaged by salt accumulation, a third of all the irrigated land in that country.

As plants take water from the soil they leave behind salt. Salt im-

How ridge-till cultivation is performed. The farmer tills two to four inches of soil in a six-inch band on top of the ridges of an old field (upper drawing) and soil from the ridges is mixed with crop residues to fill the depression between the ridges (middle drawing). Seeds are then planted on top of the ridges. Soil on ridges is usually warmer than soil between ridges, which encourages germination of the crop and deters weed germination between ridges. Black boxes indicate manure; black circles indicate crop seeds, and white ones indicate fertilizer.

ported into a field with irrigation water will inevitably accumulate in the root zone. For example, if the water contains 1,000 milligrams per liter of salt—a little more than the salt concentration of irrigation water in the Imperial Valley of California—in a year 10 tons of salt will accumulate for each hectare-foot of water applied (a hectare-foot is the amount of water needed to cover a hectare of land with a layer of water 1 foot high). Even if water contains only 100 milligrams per liter of salt, 0.33 ton of salt per hectare per year will accumulate. As much as 30 tons of salt per hectare per year can accumulate in the upper reaches of soil irrigated with slightly salty water. Accumulation of salt in the root zone eventually makes land useless for plant growth. The only solution is to flush the salt away.

Farmers can flush out salt from the root zone by applying excess water. This increases the salt concentration in runoff water, exacerbating the problem for farmers downstream. This is the case with the water from the Colorado River used to irrigate large areas in Arizona and California. The salinity of the Colorado River is less than 50 milligrams per liter in the headwaters in the Rocky Mountains but 825 milligrams per liter at the Imperial Dam in California, the last point in the United States where Colorado River water is diverted for irrigation. Where the river reaches Mexico, the concentration has increased to 950 milligrams per liter, making it barely suitable for irrigation or human use. (Actually, most of the increase in salinity is the result of natural causes—agriculture contributes only 37 percent.) The problem is so serious that the U.S. government has agreed to build an expensive desalinization plant to ensure that the water reaching Mexico is still usable. As of 1982, damage attributable to salinity in Colorado River irrigation water amounted to $113 million in California alone, and the same sources estimate that it will be more than $250 million by the year 2000.[14]

The other serious problem caused by irrigation is raising the watertable. A shallow watertable can affect root growth, first in native trees and shrubs, which in semidesert areas have deep roots, and eventually in crops. A further problem is that the watertable moves salt up from lower regions of the soil.

To reduce salt damage and keep the watertable from rising, farmers should try to limit their use of irrigation water. The cheapest and most common way of irrigating a field has been to flood it periodically. This inundates plants with more water than they need. Sprinklers are more efficient but still waste water. They are also more expensive. A promising alternative is drip irrigation, which uses a separate pipe for each plant. The pipe delivers just enough water to fulfill the needs of the plant. This system conserves water and reduces salinization but requires expensive equipment. It is viable today only with high-value crops such as specialty vegetables. The use of plastic sheeting to reduce soil evaporation and control weeds is another useful way to conserve water.

Unfortunately, none of these methods can prevent the soil from becoming salinized. Eventually accumulated salt must be flushed from the soil. In soil receiving more than 500 millimeters of rainfall a year, salt that has accumulated around the roots is flushed away. In dry areas, excess water must be applied periodically to leach the salt. The water has to infiltrate the ground rapidly, which it cannot do if soil has been compacted by machinery or if an impermeable clay layer exists close to the soil surface. There also must be subsurface drainage, that is, a way for water that penetrates lower soil levels to drain quickly into a water body such as a river or sea. This is not the case if the watertable is shallow.

In 1970 the United States lost much of the maize crop to leaf blight. One parental strain used in breeding hybrid maize was responsible for the susceptibility of the maize. Since breeders could easily replace the vulnerable strain with a resistant one, the pestilence did only temporary harm, although the losses to farmers that year were enormous. As the AIDS virus in human beings shows, entirely new threats can arise against which there are no known defenses. We grow millions of genetically uniform wheat, maize, rice, soybean, and potato plants. These are vulnerable to attack by new forms of fungi, viruses, and bacteria. Any pest that can penetrate the defenses of these widely grown plants will proliferate. Consider the grape vines in California. Until recently they were all grafted onto the same type of rootstock, which was resistant to the *Phylloxera* insect. For reasons yet unknown, this pest overcame the resistance, requiring the replacement of all the grape vines in California's Napa and Sonoma valleys. The cost has run into the billions of dollars and is bankrupting many small wineries. Had farmers used a diversity of rootstocks, the problem they are facing would have been less severe.

The reason all grape vine rootstocks were identical in California is that this particular variety throve in that region. Today's high-yield varieties are genetically more uniform than the land races they have replaced. Plant breeders who want to produce crop plants superior to traditional varieties in yield, disease resistance, and adaptability to environmental conditions need to work with as broad a spectrum of genes as possible. Ironically, the major sources of genetic diversity are the traditional varieties breeders are trying to replace. The more successful they are in producing superior varieties, the more they contribute to the decline of genetic diversity. Plant breeders find themselves between Scylla and Charybdis: If they don't continue producing superior varieties there might not be enough food for everybody, but if they continue their efforts they might run out of the raw material they need to work with. It is now clear that loss of genetic diversity, or genetic erosion, and the disappearance of wild relatives threaten the future evolution of crops.[15] The Greek wheat crop is an indication of why we should be concerned. In 1930, 80 percent of that crop came from traditional varieties; by 1965, traditional varieties contributed less than 10 percent.[16]

To solve the problem, an international program of collection and preservation of genetic diversity is under way. The program follows several steps. The first is to collect the greatest assortment possible of traditional crop varieties and wild species. Specialists then identify and catalogue the material, which must be kept alive. This is done by storing seeds or other propagules in special low-temperature environments or by growing plants in orchards, botanical gardens, agricultural experiment stations, or natural reserves. Finally, the material is distributed. The International Board of Plant Genetic Resources (IBPGR), headquartered in Rome, coordinates this vast enterprise. IBPGR is supported by the FAO and ICGAR. In addition, many countries, including the United States, have created gene banks to preserve the diversity of crops. Hundreds of thousands of crop varieties have been preserved in this manner. There are more than 400,000 samples of wheat seeds, 200,000 of rice, 100,000 of maize, beans, and soybeans, 50,000 of potatoes, 30,000 of tomatoes and cotton, and so on.

Despite these efforts, genetic erosion is a critical problem that has not yet been solved. Many crop varieties disappear before there is an opportunity to save them. Even when collectors obtain specimens, they represent only a small range of the genetic diversity of a variety. Additional diversity is lost because in each cycle of growing and sampling, only some seeds can be preserved, and because gene banks can fail for reasons varying from faulty temperature-regulation equipment to gross neglect.

Finally, it is not yet clear who owns this material. Does a country have the right to assert ownership over material it has collected in other countries? For example, can the United States claim ownership of wheat and maize varieties grown originally in Afghanistan or Mexico but preserved in U.S. gene banks? Does a country where traditional varieties grow have the right to exclude collectors from its territory? Can a country invoke political considerations in denying use of genetic resources stored in its gene banks to another? Can private parties claim exclusive rights to genetic resources?

As genetic material becomes increasingly valuable, such questions are gaining in urgency. One position asserts that materials produced by nature belong to humankind and should be available to any bona fide investigator, without restriction. The opposite view asserts that as the pool of genes diminishes and their value increases, only the market can decide their true value. In other words, they should be treated like any other valuable commodity.

Gene banks can hold only part of the genetic variation of crops. Since annual crops exist only because of human activity, gene banks are the most effective way of maintaining varieties no longer in cultivation. However, for perennial crops and the wild relatives of crops, a preferred way to conserve is by growing them in their natural habitats, what is called in-situ conservation. This requires setting aside substantial parcels of land as parks or natural reserves.

When a crop is harvested and exported to the city, it takes with it many of the nutrients that were in the soil. A small portion of these nutrients is used by people to build their bodies, but most of them end up in sewage. Ideally sewage should be recycled to the land. Unfortunately, city sewage also contains industrial waste rich in toxic compounds. Returning untreated city sewage to the land is therefore not possible. Efforts to produce organic fertilizer from sewage should be strongly encouraged.

From the earliest days of agriculture farmers have struggled to find effective ways to restore fertility to the land. Crop rotation, legumes, and animal manure were used for millennia. When scientists learned the chemical makeup of compounds that plants needed, it became possible to manufacture fertilizer.

Unfortunately, plants do not use all the fertilizer applied to soil. A significant quantity seeps into the ground and contaminates aquifers, rivers, lakes, and other bodies of water. Particularly serious is contamination by nitrates, the most commonly applied fertilizer. Excessive application of fertilizer containing phosphorus and zinc can also build up acid and aluminum in soil.

To cope with this problem, ways must be found to reduce fertilizer use without reducing yields. Better soil management will go a long way toward that goal. Another solution is a return to the ancient practice of crop rotation, abandoned in many areas with the introduction of fertilizer. It has been shown that crop rotation increases yields and reduces the need for additional fertilizer. Agronomists also recommend organic fertilizers and the most careful application of chemical fertilizers.

The loss of agricultural production to weeds, pests, and disease is sizable. Losses occur both during growing and after the harvest. Experts have estimated that yearly preharvest losses are the equivalent of 1 billion tons of food, feed, and fiber, with a combined value of about $100 billion. Close to 40 percent of agricultural production is lost to various pests after the harvest. For example, each bandicoot rat in a Bangkok rice warehouse eats about 10 kilograms of rice per year—enough to feed a human family for a week.[17] Rats eat about a third of all grain produced in Asia. Ways to eliminate or reduce these losses would aid immensely in solving the world's food problem. The technology used today depends heavily on the use of toxic chemical products. These are not very successful for a number of reasons, primarily because pest populations acquire resistance to pesticides. Chemicals have many environmentally detrimental side effects that add to their undesirability.

When first introduced, manufactured pesticides allowed farmers to cope much better with weeds and agricultural pests. Chemical herbicides eliminated the need for hand or mechanical hoeing. But the widespread use of biocides, as these synthetic products are known, has led to resistance in target species while reducing populations of other animals, including birds and mammals. A recent example is provided by the common cotton aphid, *Aphis gossypi*, in California.[18] This insect, which used to be a minor nuisance, is now a serious pest of cotton fields in the San Joaquin Valley in California. It deposits a sticky film on the mature cotton bolls, which makes the cotton sticky and can gum up ginning and yarn-spinning equipment. Wasps and other predatory insects naturally control the aphids, and it used to be only at the end of the season that they became a serious menace. For more than twenty years, farmers in California have applied insecticides during the entire cotton-growing season, especially organophosphates, to combat a variety of insects such as spider mites, thrips, whiteflies, and worms. Because of this indiscriminant application, cotton aphids have developed resistance to the insecticides while their natural enemies have been devastated. No longer checked by natural enemies, the population of cotton aphids has exploded, creating a serious problem to growers where there was only a small one before.

Pesticide residues in drinking water are an additional problem.[19] Scientists have estimated that in the United States alone, 40 percent of water supplies are contaminated with pesticide residues. So far the levels are low, but if they continue to accumulate, the risk to human health could be considerable. In much of the developing world where pesticides are used, groundwater has not been analyzed for pesticide residues.

Pesticides are a direct health hazard to farm workers. Peasants, out of ignorance or neglect, often handle toxic compounds with bare hands and take no precautions against inhaling them. Many fatal accidents are reported every year; many more go unreported. There are numerous documented cases of the long-range ill effects of certain pesticides on humans.

The use of chemical pesticides has not significantly diminished crop losses, yet agriculture, especially the high-yield type, has become dependent on them. To reduce reliance on biocides, a promising strategy based on ecological principles has been developed. It is called integrated pest management (IPM).

The primary goal of IPM is to increase plant protection while reducing the use of biocides. Under IPM farmers use resistant varieties, crop rotation, and biological control as the first lines of attack. They reserve chemical intervention for times and places when they expect a severe pest outbreak. When farmers consider it necessary to use pesticides, they apply several at a time to reduce the opportunity for evolution of resistance. Furthermore, they rotate crops to reduce the population of pathogens and avoid planting large surfaces to one single crop, intercropping two or more when possible. The thinking is that under normal circumstances, crops can be protected better if farmers allow the natural enemies of crop pests to do the controlling for them. IPM strives to encourage the growth of beneficial insects and birds and discourage the growth of pests.

IPM draws on a diversity of approaches to assess the potential threat posed by a pest and to determine the most effective and least costly way of combatting it. It is most effective when applied over a large area. This requires collaboration between all the farmers of a region, which in fact was practiced before the introduction of synthetic pesticides.

The first step in IPM is for farmers to assess how much damage they can tolerate. That is, there is a certain threshold of damage below which the costs of combatting pestilence are greater than the benefits derived from its eradication. Economic thresholds are never fixed. They depend on the market value of the crop, the effectiveness of a treatment, the climate, and so forth. IPM also requires close monitoring of potential pests. Data on pest outbreaks is fed into computer models to obtain predictions. Only when a pest population reaches a certain threshold does the farmer take measures to eliminate it.

IPM tries to defeat pests with natural controls, leaving pesticides as the last line of defense. Controls include resistant varieties, natural enemies, accurate weather predictions, and exotic species of predators. For example, the use of naturally resistant varieties reduces the pest's supply of food, while natural and introduced enemies check its propagation. Agronomists have also tried innovative techniques such as the release of millions of specially grown sterile males to combat the screwworm in southern Texas. Females mate with them and fail to reproduce, reducing the size of the natural population. Bacteria and viruses can be used to combat pests. Mangoes are susceptible to a fungal disease known as anthracnose that limits the ability of many tropical countries to export this delicacy. Plant pathologists have obtained significant control of the disease by dipping mangoes in a suspension of bacterium soon after picking. The bacterium attacks and kills the fungus before it can damage the fruit. Another technique is to have pathogenic fungi control weeds, in other words, pitting one pest against another! For example, a serious weed in many tropical crops is *Rottboelia cochinensis*. By spraying it with a solution of spores of the fungus *Colletotrichium*, farmers can substantially reduce the number of weed plants with no toxic effects.

IPM is a promising field. It requires a minimum of institutional infrastructure to monitor weather and pests, assess economic thresholds, and develop and run computer models, although detailed knowledge of a region's soil, climate, crops, and pests is essential. IPM is as good or better than chemical control, and it reduces the threat to environmental and human health. It increases biodiversity and makes farmers more aware of the ecological interaction between their crops and the rest of nature. Its use worldwide should be strongly encouraged.

Agriculture is a sector of the economy watched closely by governments. Nations require a steady supply of food to feed the nonfarming sector of the population, which today is the majority of population in all but the poorest of countries. Because food is so essential, price fluctuations can have disruptive social and economic effects. Governments take numerous measures to maintain a stable food supply, ranging from setting sanitation standards to fixing price levels. Many such measures, especially various subsidies, distort markets and adversely affect land use and productivity.[20] For example, when the government sets maximum prices for agricultural commodities to protect urban populations, the potential profit of land goes down and its value decreases. When that happens, farmers reduce their investment. They stop building structures to reduce erosion, use cheaper but less appropriate farming techniques, and buy less machinery. Less investment in the long run aggravates land degradation. When the government sets minimum prices to maintain supply or the farmer's standard of living, it encourages the overuse of land and inputs, especially fertilizer, and the farming of marginal lands. Such policies are regularly practiced in the United States and especially the European Economic Community.[21]

To protect home markets and strengthen local producers, governments have set many trade regulations. Some are intended to maintain reasonable health and sanitation standards. For example, agricultural products from areas infected with certain diseases and pests, such as hoof and mouth disease in cattle or the Mediterranean fly in produce, are not allowed into the United States. Many of these measures are reasonable, but sometimes governments use them to protect local markets, for instance by setting sanitation standards unreasonably high. In other cases, governments set high import tariffs on agricultural products, which price them out of the market. In still other cases, regulations ban imports outright to protect local growers. Japan does not import rice, for example.

Tariffs and trade barriers have the effect of devaluing the land of excluded producers, who then reduce their investment and increase the likelihood of land degradation. In turn, tariffs and trade barriers raise the value of the land of protected farmers, who are tempted to utilize marginal land and overuse good land. Because in industrialized countries agriculture contributes less to the gross domestic product, people are affected less by protectionist measures than are inhabitants of developing countries. And because farmers are politically powerful in industrialized countries they often influence their governments to institute protectionist measures. To the extent that these measures reduce wealth in the developing world, they are socially counterproductive, for third world poverty causes political and social unrest and leads to massive migration to developing countries. Refugees are costly to industrialized countries. Reducing or eliminating market distortions would mitigate the problem of displacement and encourage more rational use of land.

Agriculture today is a complex system of production, distribution, research, and marketing.[22] Crops and farming methods are constantly changing according to local and international market conditions, the introduction of new varieties, and the development of new farming techniques. All of this is supported by an extensive framework of marketing and financing, regulatory, and educational and research institutions.

Today's industrial farmers require the support of a network of marketing institutions. These provide the opportunity to meet buyers, obtain information on market prices, borrow money, insure equipment, and so forth. Where institutional support does not exist, farmers often sell their crops at low prices before the harvest to obtain the money to tide them over until the harvest.

Regulation of markets, weights and measures, sanitation, banking practices, and law enforcement are essential for the orderly development of industrial agriculture. Unfortunately, regulation is not universal. In many parts of the world, large landlords or local chieftains keep for themselves part of the farmers' legitimate earnings. This perpetuates rural poverty and thus land degradation through overuse and underinvestment.

Modern farming also depends on research and development to improve crop varieties, farming practices, machinery, and sanitation methods. In most countries, research and development is conducted at universities and agricultural research institutions. They train farmers and maintain a network of regional consultants who keep in touch with farmers. Plant-breeding companies and agrochemical and farm-machinery manufacturers send out trained salespersons to help diffuse the latest technologies. Farmers who have access to these networks of information are clearly at an advantage. Many farmers in the developing world are not sustained by research and education networks and therefore cannot benefit quickly from technical advances. One way to increase productivity in the next fifty years is to create such institutional support throughout the world.

Finally, farmers require a system of infrastructure including roads, electricity, communications, health services, and retail services to maintain a reasonable standard of living. A fifty-mile road that serves ten farms is less visible than a five-mile paved city road that serves several thousand persons, and governments with limited resources often neglect rural infrastructure. Living and working on an undersupported farm can mean deprivation. Thus the most prosperous farmers or their children often exchange rural for city life. This perpetuates the vicious circle of rural poverty, ignorance, tradition, and underinvestment, which at best keeps production low and at worst leads to land degradation. Good farming requires well-educated farmers. To retain them, the standard of living on the farm must be comparable to that in the city.

A serious impediment to more efficient farm production, which we will only mention here, is inequitable distribution of land. In many countries, 10 percent of farmers own 80 percent of the land, and there are regions where the ratio is even more inequitable. Concentration of land in a few hands is inefficient and environmentally damaging. Owners of large surfaces tend to exploit their holdings extensively for lack of capital (because all their capital is in the land). Those who have too little land undercapitalize their efforts. On small plots, for example, savings from modern machinery are not sufficient to justify the cost of the machinery. Furthermore, small landholders overuse their land out of necessity. [23]

The second half of the twentieth century has sometimes been called the era of biology on account of the unprecedented growth it saw in biological knowledge. Several factors brought about this growth. The discovery of the biological role of DNA and the unraveling of its structure opened up the cell to the investigation of physics and chemistry.

A massive influx of public funds for research and the efforts of thousands of biologists increased understanding at least a hundredfold. These changes were fueled by the introduction of computers and advances in scientific instrumentation.

Research has greatly expanded understanding of the performance of organisms. Knowledge is so detailed that scientists can now create plants and animals with unique characteristics. Organisms can be genetically programmed to do things that no wild organism can. For example, scientists have created bacteria that produce human insulin. In principle bacteria can be made to produce any drug that plants or animals produce naturally, such as caffeine or nicotine. As of now, however, the promise of biotechnology exceeds it successes.

Plant biotechnology involves two methodologies: genetic engineering and in vitro production of plants from single cells. Genetic engineering refers to a series of techniques that allow plant biologists to insert genes from one species into another. With in vitro production, biologists can take a single plant cell, let it multiply in a culture, and by treating that culture with a sequence of plant hormones, grow an entire plant. This technique, now in commercial use, is how most new varieties of horticultural species are produced. It ensures that plants are genetically identical, which is important in the trade but, as discussed above, has serious drawbacks.

Plant breeders have been introducing genes from one variety or species into another through hybridization and selection since time immemorial. Biotechnology has two advantages over these conventional forms of breeding. First, the process of crossing and selection takes a minimum of ten generations to produce a true breeding variety; new genes can be introduced through genetic engineering in only one generation. Second, the conventional plant breeder introduces genes only from species that can interbreed, while the genetic engineer is not so constrained.

Biologists hope to introduce into crop cells genes for resistance to pests, spoilage, frost, and heat; for nutritional quality; and for fast growth to increase the value of the crop and its productivity. The idea is to multiply these cells and produce hundreds of thousands of plants to order. Scientists also hope to produce crops that yield valuable compounds obtained until now through expensive chemical or biochemical processes. The principal application of this knowledge so far has been in the fermentation and pharmaceutical industries.

Biotechnology has the potential to solve many problems in medicine, agriculture, and ecology. In the medical area, it will allow the creation of new pharmaceutical compounds. Defective human genes might be replaced with normal ones. Compounds such as human insulin will be readily synthesized by specially engineered organisms. In agriculture, biotechnology expects to produce plants specially suited for local growing conditions, with proteins that have amino acids balanced for human nutrition. Besides developing varieties resistant to common pests, which would reduce the need for pesticides, biotechnologists are trying to make cereal varieties that can fix nitrogen from the air (as legumes do), which would reduce the need for chemical fertilizer. In the environmental area, biotechnologists are working on bacteria that consume petroleum products and could be sown over oil spills. They also are trying to develop microorganisms that can reduce toxic chemicals to harmless compounds. Little of this promise has been realized, and it will probably not be until the twenty-first century that biotechnology comes into its own.

The oil palm, *Elais guinensis*, provides a glimpse of what may be in store. The need for palm oil for human consumption and soap production has soared with the growth of world population over the last thirty years. The oil palm is native to western Africa but cultivated in many parts of the tropics. Palms have only one trunk, and the leaves and fruit are produced at the end of the stem. When the palms get too tall, harvesting their fruit becomes expensive. This means that farmers replace plantation trees approximately every twenty-five years. Plantations are reproduced from seeds, but because the species is variable and cannot self-pollinate, it is difficult to obtain uniform populations and develop improvement programs. However, using traditional breeding methods, the Institut des Recherches sur les Huiles et les Oleagineaux in the Ivory Coast increased yields from 1 metric ton per hectare in 1960 to about 4 metric tons per hectare in 1980.

From 1970 on, French and English biotechnologists embarked on a program to clone the oil palm using tissue culture techniques. By selecting cells from the highest-yield plants, they boosted yields over 20 percent in one generation.[24] An impressive amount of basic research supported this investigation. Scientists can further improve yields by the introduction of genetic engineering techniques that promise to control fatty-acid composition, increase pest resistance, and change plant structure to reduce harvesting costs.

Another example of the successful application of biotechnology is resistance to viral attack, the cause of serious crop losses every year. Investigators learned that cells containing genes that turn a virus's coat into protein were strongly resistant to attack by the virus. Using this knowledge, researchers have now engineered a range of crops with resistance to a dozen different plant viruses.[25]

An ingenious procedure to make species produce their own insecticide involves the transfer of genes from a bacterium, *Bacillus thuringiensis*, to crops. This bacterium is poisonous if ingested by the caterpillars of moths and butterflies but harmless to other insects and to vertebrates. For some time, farmers have sprayed preparations of the bacterium on the leaves of crops. However, the bacterium is easily washed away by rain and is effective for only a short time. Biotechnologists isolated the bacterium genes that make the poisonous compound and introduced them into cotton plants, producing varieties permanently resistant to all caterpillar pests, including the bollworm.

Another example of the application of biotechnology to agriculture is the introduction of spoilage resistance into tomatoes. Every year millions of dollars' worth of ripe produce spoils because of premature ripening. Most fruits ripen as a result of the natural production of the gas ethylene, controlled by a specific gene. Tomato plants engineered to suppress the gene for ethylene production ripen only if ethylene is applied externally. That operation can now take place just before they go on sale.[26]

Because biotechnology research is expensive, requiring sophisticated instruments and highly skilled labor, most but not all of it is being done in the developed world. Applying biotechnology techniques is also expensive. Costs will probably come down as work moves from the experimental to the operational phase. Yet capital outlays will still be high and well beyond the reach of the average farmer. Patents on the new miracle breeds will further drive up costs and limit access.

Creating a biotechnology industry requires large capital expenditures. Venture capitalists and multinational corporations are willing to invest in this promising but risky industry only if they can be assured of a fair return on their investment. In response to this problem, the United States, Europe, and Japan have passed laws to protect developers of organisms with artificially created characteristics. Patents give exclusive rights to their creators. The situation is analogous to that of an industrial product patented by Dupont. With its monopoly for many years on the production of nylon, Dupont was able to recover its original investment. Patents are an incentive to invest in biotechnology research and development. Yet the system may have negative implications for agriculture.

Although today's networks of agricultural experiment stations that produce new varieties and improve farming methods require capital investment and specialized labor, all but the poorest of nations can afford them. The free transfer of information and genetic stocks greatly eased their establishment. By contrast, in the field of biotechnology, where procedures and organisms will be the property of their developers, a few multinational corporations are likely to have a near monopoly on new organisms.[27] Unless each country can develop its own biotechnology industry, improvements in agriculture will become dependent on seeds commercially produced elsewhere. Because seeds are proprietary, presumably firms will not sell them to farmers. Instead, farmers will have to be licensed and pay a royalty each time they sow. Entire countries could become dependent on seeds produced outside their borders to feed their populations.

If we apply biotechnology without thinking through the social and economic consequences, we could find ourselves impoverishing humanity rather than enriching it. If biotechnology requires capital investments available only to the few, and if the seeds of miracle plants are not available to everybody, then only a few will benefit. The rest of humanity will be pushed further into poverty and despair. The promise of biotechnology must not become a tool for exploitation.

The potential impact on agriculture of actual and predicted human-made climatic and environmental changes, referred to collectively as global change, is enormous.[28] Climate change is the result of both natural and human-made forces. But through changes in land use such as removal of vegetation, soil degradation, and draining of marshes, and through the release of pollutants into the water and the air, humans are affecting the dynamics of atmospheric physical and chemical processes as never before in their history.

Climate is a complex process, and thus potential changes in its patterns are difficult to predict. Some changes have already happened, however. For example, chemical compounds, particularly chlorofluorocarbons, that combine with ozone have reduced the stratospheric ozone layer. The concentration of carbon dioxide in the atmosphere has almost doubled in the last 150 years, and the concentration of other greenhouse gases such as methane has increased significantly. These compounds are all by-products of industrial and agricultural activities, especially the burning of fossil fuels. Scientists have predicted that the increase in greenhouse gases will warm the atmosphere by an average of 1 to 3° C in the next 100 to 150 years.[29] Such a change would alter winter and summer maximum temperatures and bring shifts in the distribution of crops. It would also alter rainfall distribution, a potentially devastating prospect. According to one scenario, the rich farmlands of the midwestern United States would get warmer and drier, significantly lessening cereal yields, especially wheat and maize. Another dim prediction is that the increase in ultraviolet radiation resulting from a thinner ozone layer will augment the mutation rate in plants and animals and reduce crop production.

Global change could alter the delicate balance of industrial agriculture enough that it will not meet its target of a minimum of 2 percent annual growth for the next hundred years. Because of serious problems related to soil, water, genetics, markets, and institutional frameworks, meeting that target will be difficult even without large climatic changes. It will be close to impossible if these changes happen. Humanity could plunge into a serious food crisis, with unforeseen consequences. We should make every effort to reduce the likelihood of climatic change, or at least to keep it to a minimum. This entails lowering emissions of carbon dioxide, especially from cars, banning the use of chlorofluorocarbons and other ozone-reducing chemicals, and adopting agricultural and industrial practices to mitigate or eliminate land degradation.

When people adopted agriculture, they traded the security of a large variety of plant food and game for the higher yield of a few superior plants. Increased food production made it possible to feed a larger population. Food surpluses gave birth to a leisure class of rulers, priests, artists, poets, and philosophers. It also gave birth to inequality and struggles for land and power.

Over the ages, this increased appropriation of the produce of the earth has required a greater and greater investment of human labor, supplemented first by animal and then by fossil energy. The wild wheat, barley, rice, sorghum, maize, manioc, and potatoes domesticated by our Neolithic ancestors originally produced only a fraction of what their descendants produce today. As human populations grew and needed more food, people learned to direct the evolution of their crops so that they yielded more. Purposeful manipulation of crops and experimentation with farming techniques followed. With the advent of plant biology and especially genetics, what was once haphazard selection has become highly directed and systematic. It has greatly expanded yield but also restricted the genetic variability of most crops. Knowledge of plant nutrition and soil chemistry led to inorganic fertilizers and further advances in production. The latest step in appropriating the produce of the earth is biotechnology: the industry of producing plants to order. With it the relation of the farmer and of society to the natural ecosystem becomes more remote than ever.

Today the by-products of human civilization are threatening the very things that made civilization possible. By multiplying so tremendously over the last one hundred years and augmenting industrial activity, people have increased their demands on nature severalfold. The need of billions of people for land and water is endangering thousands of species of plants, animals, and microorganisms.[30] In addition, the use of fossil fuels in industry and transportation, compounded by the release of innumerable other chemical products slowly poisoning soil, water, and air, threatens to change the dynamics of climate.

Until recently, most of humankind saw nature as created to serve people.[31] The notion that humans have the responsibility of conserving for future generations natural landscapes and the plants and animals that inhabit them is quite recent. The dramatic environmental changes we are witnessing today have created this fundamental change in view from humankind as conqueror to humankind as shepherd of nature.

How exactly agriculture will meet the challenge of increasing worldwide production by 2 to 3 percent per annum for 150 years is difficult to predict. This much is certain: Agriculture of the future will be even more technical. Inputs of fertilizer and biocide will be high, but less per unit output than today.

Farmers, especially in developing countries, will need to be much more educated, and farms will require more in the way of capital investment. Most likely specialized corporations will replace individual family farms, as corporate factories replaced family shops during the industrial revolution. The proportion of people living in the country will steadily decline. Of those employed in farming, a large number will be salaried employees and only a few will be landowners. Farming as the world has known it for the last twenty centuries will disappear. Plants will be radically altered until none but the scientist will know the connection between them and their wild ancestors. And yet none of these developments will alter the fact that of all the activities ever invented by humankind, agriculture is the most basic to its survival.

Notes

PROLOGUE

1. "Una cosa deseo ver acabada de tratar, y es lo que toca a la conservación de los bosques y aumento de ellos que es mucho menester y creo que andan muy al cabo; temo que los que vinieren después de nosotros han de tener mucha queja de que se los dejamos consumidos, y pliegue a Dios que no lo veamos en nuestros días" (free translation by the authors).

2. H. G. and F. G. A. Wolman, *Land Transformation in Agriculture* (Chichester, U.K.: John Wiley, 1987).

3. M. Rothschild, *Bionomics: The Inevitability of Capitalism* (New York: Henry Holt, 1990).

4. M. Oelschlaeger, *The Idea of Wilderness* (New Haven: Yale University Press, 1991).

5. World Bank, *World Development Report, 1991: The Challenge of Development* (Oxford: Oxford University Press, 1991).

6. *Economic Report of the President* (Washington, D.C.: GPO, 1981).

7. P. L. Pellet, "Problems and Pitfalls in the Assessment of Human Nutritional Status," in M. Harris and E. B. Ross, eds., *Food and Evolution* (Philadelphia: Temple University Press, 1987).

8. A. W. Crosby, *Ecological Imperialism: The Biological Expansion of Europe, 900–1900* (Cambridge, U.K.: Cambridge University Press, 1986).

9. F. Di Castri, "History of Biological Invasions with Special Emphasis on the Old World," in J. A. Drake *et al.*, eds., *Biological Invasions: A Global Perspective* (Chichester, U.K.: John Wiley, 1989).

10. R. L. Brown, *In the Human Interest* (New York: W. W. Norton, 1974).

11. World Resources Institute, *World Resources, 1988–89* (New York: Basic Books, 1989).

12. R. Dawkins, *The Blind Watchmaker* (New York: W. W. Norton, 1987).

13. E. Boserup, *The Conditions of Agricultural Growth* (London: Allen and Unwin, 1981).

CHAPTER 1

1. B. M. Fagan, *People of the Earth: An Introduction to World Prehistory* (Boston: Little, Brown, 1986).

2. B. Campbell, *Human Evolution: An Introduction to Man's Adaptation*, 3d ed. (New York: Aldine, 1985).

3. R. Foley, *Another Unique Species: Patterns in Human Evolutionary Ecology* (Harlow: Longman Scientific and Technical, 1987).

4. P. E. Martin, "Prehistoric Overkill," in P. S. Martin and H. E. Wright, eds., *Pleistocene Extinctions: The Search for a Cause* (New Haven: Yale University Press, 1967).

5. M. N. Cohen, "The Significance of Long-Term Changes in Human Diet and Food Economy," in M. Harris and E. B. Ross, eds., *Food and Evolution* (Philadelphia: Temple University Press, 1987).

6. B. Campbell, *Human Evolution.*

7. M. D. Sahlins, *Stone Age Economics* (Chicago: Chicago University Press, 1972).

8. B. Meehan, "Hunters by the Seashore," *Journal of Human Evolution* 6: 363–70; K. Hill and A. M. Hurtado, "Hunter Gatherers of the New World," *American Scientist* 77: 437–43.

9. J. H. Steward and L. C. Faron, *Native Peoples of South America* (New York: McGraw-Hill, 1959).

10. R. B. Lee and E. DeVore, eds., *Man the Hunter* (Chicago: Aldine, 1968).

11. H. Kaplan *et al.*, "Food Sharing among the Ache Hunter-Gatherers of Eastern Paraguay," *Current Anthropology* 25: 113–15.

12. M. Sahlins, "Notes on the Original Affluent Society," in Lee and DeVore, *Man the Hunter*.

13. F. Parkman, Jr., *The Oregon Trail* (Harmondsworth, U.K.: Penguin Books, 1982).

14. Lee and DeVore, *Man the Hunter*.

15. E. Yanovsky, *Food Plants of the North American Indians* (Washington, D.C.: USDA, miscellaneous publication, 1936).

16. E. B. Ross, "An Overview of Trends in Dietary Variation from Hunter-Gatherer to Modern Capitalist Societies," in Harris and Ross, *Food and Evolution*.

17. M. N. Cohen, *The Food Crisis in Prehistory: Overpopulation and the Origins of Agriculture* (New Haven: Yale University Press, 1977); M. N. Cohen and G. J. Armelagos, eds., *Paleopathology at the Origins of Agriculture* (Orlando: Academic Press, 1984).

18. S. C. Watkins and E. van de Walle, "Nutrition, Mortality, and Population Size: Malthus' Court of Last Resort," in R. I. Rotberg and T. K. Rabb, eds., *Hunger and History: The Impact of Changing Food Production and Consumption Patterns on Society* (Cambridge, U.K.: Cambridge University Press, 1983).

19. V. Johnson, "The Causes of Hunger," *Food and Nutrition Bulletin (UNU/ WHP)* 3 (2): 1–9; M. Wolde-Marian, *Rural Vulnerability to Famine in Ethiopia, 1958–77* (New Delhi: Vikas, 1984).

CHAPTER 2

1. R. B. Lee and E. DeVore, eds., *Man the Hunter* (Chicago: Aldine, 1968).

2. R. S. MacNeish, *The Origins of Agriculture and Settled Life* (Norman, Oklahoma: University of Oklahoma Press, 1991).

3. M. N. Cohen, *The Food Crisis in Prehistory: Overpopulation and the Origins of Agriculture* (New Haven: Yale University Press, 1977).

4. R. J. Braidwood *et al.*, *Prehistoric Investigations in Iraqui Kurdistan* (Chicago: University of Chicago, 1960).

5. H. E. Wright, Jr., "Natural Environment of Early Food Production North of Mesopotamia," *Science* 161: 334–39; W. Van Zeist, "Reflections on Prehistoric Environments in the Near East," in P. Ucko and G. W. Dimbleby, eds., *The Domestication and Exploitation of Plants and Animals* (Chicago: Aldine, 1969).

6. V. G. Childe, *Piecing Together the Past* (London: Routledge and Kegan Paul, 1956).

7. E. Anderson, *Plants, Man, and Life* (Boston: Little, Brown, 1952); J. G. Hawkes, *The Diversity of Crop Plants* (Cambridge: Harvard University Press, 1983).

8. T. H. Engelbrecht, "Uber die Entstehung einiger Feldmassig Angebauter Kulturpflanzen," in *Geographische Zeitshrift* 22: 328–35; Anderson, *Plants, Man, and Life*; C. O. Sauer, *Agricultural Origins and Dispersal* (Cambridge, Massachusetts: MIT Press, 1969).

9. B. Bender, "Gatherer-Hunter to Farmer: A Social Perspective," *World Archeology* 10: 204–22; B. Hayden, "Nimrods, Piscators, Pluckers, and Planters: The Emergence of Food Production," *Journal of Anthropological Research* 9: 31–69.

10. M. W. Prausnitz, *From Hunter to Farmer and Trader* (Jerusalem, 1970); R. S. MacNeish, "The Beginning of Agriculture in Central Peru," in C. A. Reed, ed., *Origins of Agriculture* (The Hague: Mouton, 1977).

11. Botanically speaking, "seeds" of grasses are fruits of the type known as caryopses, or grains. We will use "seeds" in the sense it carries outside of scientific circles.

12. J. Mellaart, *The Neolithic of the Near East* (London: Thames and Hudson, 1975); C. Burney, *From Village to Empire* (Oxford: Phaidon, 1977).

13. B. Orme, "The Advantages of Agriculture," in J. V. S. Megaw, ed., *Hunters, Gatherers, and First Farmers beyond Europe* (Leicester, U.K.: Leicester University Press, 1977).

14. Lee and DeVore, *Man the Hunter*; M. D. Sahlins, *Stone Age Economics* (Chicago: Chicago University Press, 1972).

15. Cohen, *The Food Crisis*.

16. W. H. McNeill, *The Rise of the West* (Chicago: University of Chicago Press, 1963).

17. N. I. Vavilov, "Studies on the Origin of Cultivated Plants," *Bulletin of Applied Botany, Genetics and Plant Breeding* 16: 1–248 (1926, in Russian); *Estudio sobre el Origin de las Plantas Cultivadas* (Buenos Aires: Acme, Spanish translation).

18. G. P. Murdock, *Africa: Its People and Their Cultural History* (New York, 1959).

19. J. R. Harlan, "Agricultural Origins: Centers and Non Centers," *Science* 174: 465–73 (1971); R. S. MacNeish, *The Origins of Agriculture*.

20. D. Rindos, *The Origins of Agriculture* (Orlando, Florida: Academic Press, 1984).

21. E. S. Deevey, "The Human Population," *Scientific American* 203: 195–200 (1960).

22. D. Grigg, *The Dynamics of Agricultural Change* (London: Hutchinson, 1982).

23. J. R. Kloppenburg, Jr., *First the Seed* (Cambridge, U.K.: Cambridge University Press, 1988).

CHAPTER 3

1. W. Van Zeist, "Reflections on Prehistoric Environments in the Near East," in P. Ucko and G. W. Dimbleby, eds., *The Domestication and Exploitation of Plants and Animals* (Chicago: Aldine, 1969).

2. O. Bar-Yosef and A. Belfer-Cohen, "From Foraging to Farming in the Mediterranean Levant," in A. B. Gebauer and T. D. Price, *Transitions to Agriculture in Prehistory* (Madison, Wisconsin: Prehistory Press, 1992).

3. A. T. Moore, "The Development of Neolithic Societies in the Near East," *Advances in World Archeology* 4: 1–69 (1985).

4. F. Hole, K. Flannery, and J. Neely, "Early Agriculture and Animal Husbandry in Deh Luran, Iran," *Current Anthropology* 6: 105–6 (1965).

5. J. R. Harlan and D. Zohary, "Distribution of Wild Wheats and Barley," *Science* 153: 1075–80 (1966).

6. J. Harlan, "A Wild Wheat Harvest in Turkey," *Archeology* 20: 197–201 (1967).

7. E. R. Sears, "Wheat Cytogenetics," *Annual Review of Genetics* 3: 451–68 (1969). See also G. Kimber, "The Relationships of the S-Genome Diploids to Polyploid Wheats," *Proceedings 4th International Wheat Genetics Symposium* 6: 81–85 (1974).

8. M. Feldman, "Wheats," in N. W. Simmonds, *Evolution of Crop Plants* (London: Longman, 1976).

9. F. Braudel, *Civilization and Capitalism, 15th–18th Century*, vol. 1, *The Structures of Everyday Life* (New York: Harper and Row, 1981).

10. G. Hillmann, "On the Origin of Domesticated Rye, *Secale Cereale*: The Finds from Aceramic Can Hasan III in Turkey," *Anatolian Studies* 28: 157–74 (1978).

11. J. Mellaart, *Catal Huyuk: A Neolithic Town in Anatolia* (London: Thames and Hudson, 1967); M. Grant, *The Ancient Mediterranean* (New York: Penguin, 1969).

12. Vishnu-Mire, "Changing Economy in Ancient India," in C. A. Reed, ed., *Origins of Agriculture* (The Hague: Mouton, 1977).

13. P. T. Ho, "The Loess and the Origin of Chinese Agriculture," *American Historical Review* 75: 1–36 (1969).

14. E. N. Anderson, *The Food of China* (New Haven: Yale University Press, 1988).

15. J. Needham and F. Bray, *Sciences and Civilization in China*, vol. 6 (2), *Agriculture* (Cambridge, U.K.: Cambridge University Press, 1984).

16. Anderson, *The Food of China.*

17. I. C. Glover, "Some Problems Relating to the Domestication of Rice in Asia," in V. N. Misra and P. Bellwood, eds., *Recent Advances in Indo-Pacific Prehistory* (New Delhi: Oxford and IBH, 1985); K. L. Mehra and R. K. Arora, "Some Considerations on the Domestication of Plants in India," in Misra and Bellwood, *Recent Advances.*

18. O. T. Solbrig, ed., "The Southern Andes and Sierras Pampeanas," *Mountain Research and Development* 4: 97–190 (1984).

19. B. Pickersgill and C. Heiser, "Origins and Distribution of Plants Domesticated in the New World Tropics," in Reed, *Origins of Agriculture.*

20. M. León-Portilla, "Mesoamerica before 1519," in L. Bethell, ed., *The Cambridge History of Latin America*, vol. 1 (Cambridge, U.K.: Cambridge University Press, 1984).

21. R. S. MacNeish, "Ancient Mesoamerican Civilization," *Science* 143: 531–37 (1964).

22. K. V. Flannery, "The Origins of Agriculture," *Annual Reviews of Anthropology* 2: 271–310 (1973).

23. P. C. Mangelsdorf, *Corn: Its Origin, Evolution and Improvement* (Cambridge: Harvard University Press, 1974).

24. G. Wilkes, *Teosinte: The Closest Relative of Maize* (Cambridge: Bussey Institution of Harvard University, 1967); G. Wilkes, "Maize: Domestication, Racial Evolution, and Spread," in D. R. Harris and G. C. Hillman, *Foraging and Farming* (London: Unwin Hyman, 1989).

25. L. Kaplan, "Archaeology and Domestication in American Phaseolus Beans," *Economic Botany* 19: 358–68 (1965).

26. R. S. MacNeish, "The Beginning of Agriculture in Central Peru," in Reed, *Origins of Agriculture.*

27. L. Kaplan, "Variation in Cultivated Beans," in T. F. Lynch, ed., *Guitarrero Cave* (New York: Academic Press, 1980).

28. MacNeish, *The Origins of Agriculture.*

29. D. M. Pearsall, "Pachamachay Ethnobotanical Report: Plant Utilization of a Hunting Base Camp," in J. Rick, ed., *Prehistoric Hunters of the High Andes* (New York: Academic Press, 1980); D. M. Pearsall, "Adaptation

of Early Hunter-Gatherers to the Andean Environment," in Harris and Hillman, *Foraging and Farming*.

30. A. J. St. Angelo and G. E. Mann, "Peanut Proteins," in *Peanut: Culture and Uses* (Stillwater, Oklahoma: Oklahoma University Press, 1973).

31. A. Krapovickas, "Evolution of the Genus *Arachis*," in R. Moav, ed., *Agricultural Genetics* (New York: Hammons, 1973).

32. C. O. Sauer, *Agricultural Origins and Dispersals* (New York: American Geographic Society, 1952).

33. M. B. Bush, D. R. Piperno, and P. A. Colinvaux, "A 6,000-Year History of Amazonian Maize Cultivation," *Nature* 340: 303–5 (1989).

34. N. W. Simmonds, *Evolution of Crop Plants* (London: Longman, 1976).

35. G. P. Murdock, *Africa: Its People and Their Cultural History* (New York, 1959).

CHAPTER 4

1. R. Unger-Hamilton, "The Epi-Paleolithic Southern Levant and the Origins of Cultivation," *Current Anthropology* 30: 88–104.

2. N. M. Nayar and K. I. Mehra, "Sesame: Its Uses, Botany, Cytogenetics, and Origin," *Economic Botany* 24: 20–31 (1970).

3. Self-pollination refers to the process of pollen germinating on the styles of its own plants, which may or may not result in successful fertilization. Self-fertilization, on the other hand, is the process of fertilization of the ovule by the pollen of the same plant.

4. J. Mager, M. Chevion, and G. Glaser, "Favism," in I. E. Liener, *Toxic Constituents of Plant Foodstuffs* (New York: Academic Press, 1980).

5. Principally, two cytogenetic glycosides, *linamarin* and *lotraustalin*.

6. J.-P. Aeschlimann, "The Potential of Biological and Integrated Control of Pests in Sub-Saharan Africa, and Conditions for Its Implementation," *Biology International* 17: 10–14 (1988).

7. D. Pimentel, "The Dimensions of the Pesticide Question," in F. H. Borman and S. R. Kellert, *Ecology, Economics, Ethics: The Broken Circle* (New Haven: Yale University Press, 1991).

8. E. Hernandez-Xolocotzi, *Exploraciones Ethnobotanicas y su Metodologia* (Chapingo, Mexico: Escuela Nacional de Agricultura, 1970); G. Wilkes, "Maize: Domestication, Racial Evolution and Spread," in D. R. Harris and G. C. Hillman, *Foraging and Farming* (London: Unwin Hyman, 1989).

9. Wilkes, "Maize."

10. C. B. Heiser, "Some Considerations of Early Plant Domestication," *BioScience* 19: 228–31 (1969).

11. J. R. Harlan and D. Zohary, "Distribution of Wild Wheats and Barley," *Science* 153: 1074–80 (1966).

12. D. R. Harris, "The Origins of Agriculture in the Tropics," *American Scientists* 60: 180–93 (1972); F. Braudel, *Civilization and Capitalism, 15th–18th Century*, vol. 1, *The Structures of Everyday Life* (New York: Harper and Row, 1981).

13. K. V. Flannery, "The Origins of Agriculture," *Annual Reviews of Anthropology* 2: 271–310 (1973).

14. A. Roosevelt, "The Evolution of Human Subsistence," in M. Harris and E. B. Ross, eds., *Food and Evolution: Toward a Theory of Human Food Habits* (Philadelphia: Temple University Press, 1987).

15. H. F. De Luca, "Vitamin D: The Vitamin and the Hormone," *Federation Proceedings* 33: 2211–19 (1974).

CHAPTER 5

1. From the Latin *civitas*, meaning "community of citizens," or "city."

2. M. Grant, *The Ancient Mediterranean* (New York: Meridian Books, 1988).

3. W. H. McNeill, *The Rise of the West* (Chicago: University of Chicago Press, 1963).

4. D. Zohary and M. Hopf, *Domestication of Plants in the Old World* (Oxford: Clarendon Press, 1988).

5. J. D. Clark, "The Domestication Process in Northeast Africa: Ecological Change and Adaptive Strategies," in L. Krzyzaniak and M. Kubusiewcz, eds., *Origin and Early Development of Food-Producing Cultures in North-Eastern Africa* (Poznan: Polish Academy of Sciences, 1984).

6. Herodotus, *Histories* 2, 14.

7. R. J. Forbes, *Studies in Ancient Technology*, vol. 4 (Leiden, the Netherlands: E. J. Brill, 1956).

8. F. E. Zeuner, "Cultivation of Plants," in C. J. Singer, ed., *The History of Technology*, vol. 1 (Oxford: Clarendon Press, 1954).

CHAPTER 6

1. W. H. McNeill, *The Rise of the West* (Chicago: University of Chicago Press, 1963).

2. For a discussion of a different view, see G. Barker, *Prehistoric Farming in Europe* (Cambridge, U.K.: Cambridge University Press, 1985).

3. McNeill, *The Rise of the West.*

4. R. Pittioni, "Der Urgeschichtliche Horizont der Historischen Zeit," in G. Mann and A. Heus, eds., *Propylaen Weltgeschichte*, vol. 1 (Berlin: Propylaen Verlag, 1961); J. V. Vives, *Historia de España y América, Social y Económica*, vol. 1, *Antigüedad, Alta Edad Media, América Primitiva* (Barcelona: Vivens-Vives, 1972).

5. C. Renfrew, *The Emergence of Civilization* (London: Methuen, 1976).

6. P. Halsteadt, "Counting Sheep in Neolithic and Bronze Age Greece," in I. Hodder, G. Isaac, and N. Hammond, eds., *Pattern of the Past: Studies in Memory of David Clarke* (Cambridge, U.K.: Cambridge University Press, 1981).

7. Y. V. Andreyev, "The World of Crete and Mycenae," in I. M. Diakonoff, *Early Antiquity* (Chicago: University of Chicago Press, 1991).

8. A. Evans, *The Palace of Minos* (London: Macmillan, 1921).

9. G. E. Fussell, *Farms, Farmers, and Society* (Lawrence, Kansas: Coronado Press, 1976).

10. L. Pearl, *Rice, Spice and Bitter Oranges: Mediterranean Foods and Festivals* (New York: World Publishing, 1967).

11. R. J. Forbes, *Studies in Ancient Technology*, vol. III (Leiden: E. J. Brill).

12. M. Grant. *The Ancient Mediterranean* (New York: Meridian, 1988).

13. *Ibid.*

14. M. E. Lowenberg et al., *Food and People* (New York: John Wiley, 1979).

15. V. D. Hanson, *Warfare and Agriculture in Classical Greece* (1983).

16. C. Tacitus, *Germania*, chapter 26, as quoted by Gras 1925.

17. D. Grigg, *The Dynamics of Agricultural Change* (London: Hutchinson, 1982).

18. M. Johnston, *Roman Life* (Chicago: Scott, Foresman, 1957).

19. Lowenberg, *Food and People.*

20. J.-B. Harersath, *Die Agrarlandschaft im Romischen Deutschland der Kaiserzeit* (Passau: Passavia Universitätsverlag, 1984).

21. As cited by A. Dickson, *The Husbandry of the Ancients*, 2 vols. (Edinburgh, 1787).

22. *Ibid.*

23. E. G. Lamb and E. G. Mittleberger, *In Celebration of Wine and Life* (New York: Drake, 1974).

24. V. G. Carter and T. Dale, *Topsoil and Civilization* (Norman, Oklahoma: University of Oklahoma Press, 1974).

25. M. Grenon and M. Batisse, *Futures for the Mediterranean Basin* (Oxford: Oxford University Press, 1989).

CHAPTER 7

1. N. S. B. Gras, *A History of Agriculture in Europe and North America* (New York: S. F. Crofts, 1925).

2. D. Hartley, *Lost Country Life* (New York: Pantheon, 1979).

3. R. Tannahill, *Food in History* (New York: Stein and Day, 1973).

4. C. Braudel, *Civilization and Capitalism, 15th–18th Century*, 3 vols. (New York: Harper and Row, 1981–84); J. deVries, *European Urbanization, 1500–1800* (Cambridge: Harvard University Press, 1984).

5. I. Wallerstein, *The Modern World System*, 2 vols. (New York: Academic Press, 1974).

6. A. M. Watson, *Agricultural Innovation in the Early Islamic World* (Cambridge, U.K.: Cambridge University Press, 1983).

7. N. Neilson, *Medieval Agrarian Economy* (New York: Henry Holt, 1936).

8. D. N. McCloskey, "The Persistence of English Common Fields," in W. N. Parker and E. L. Jones, eds., *European Peasants and Their Markets* (Princeton: Princeton University Press, 1975).

9. For a good and comprehensive discussion of these changes, see Braudel, *Civilization and Capitalism*. For a slightly different view, see Wallerstein, *The Modern World System*.

10. J. deVries, "Peasant Demand Patterns and Economic Development: Friesland, 1550–1750," in Parker and Jones, *European Peasants*; deVries, *European Urbanization*.

11. R. Davis, *The Rise of the Atlantic Economies* (Ithaca: Cornell University Press, 1973).

12. D. Grigg, "The World's Agricultural Labor Force, 1800–1970," *Geography* 60: 194–202 (1975).

CHAPTER 8

1. J. H. Galloway, *The Sugar Cane Industry: An Historical Geography from Its Origins to 1914* (Cambridge, U.K.: Cambridge University Press, 1989).

2. E. N. Anderson, *The Food of China* (New Haven: Yale University Press, 1988).

3. R. T. Gunther, *The Greek Herbal of Dioscorides* (New York: Hafner, 1959).

4. A. M. Watson, *Agricultural Innovation in the Early Islamic World* (Cambridge, U.K.: Cambridge University Press, 1983).

5. A. Hourani, *A History of the Arab Peoples* (Cambridge: Harvard University Press, 1991).

6. J. Lockhart and S. B. Schwartz, *Early Latin America* (Cambridge, U.K.: Cambridge University Press, 1983).

7. S. B. Schwartz, *Sugar Plantations in the Formation of Brazilian Society: Bahia, 1550–1835* (Cambridge, U.K.: Cambridge University Press, 1985). The discussion of the development of the Brazilian sugar industry is largely based on this excellent study.

8. Galloway, *The Sugar Cane Industry*.

9. H. Hobhouse, *Seeds of Change* (New York: Harper and Row, 1986).

10. M. M. Fraginals, *The Sugarmill: The Socioeconomic Complex of Sugar in Cuba, 1760–1860* (New York: Monthly Review Press, 1976).

CHAPTER 9

1. C. Braudel, *Civilization and Capitalism, 15th–18th Century*, 3 vols. (New York: Harper and Row, 1981–84); I. Wallerstein, *The Modern World System* (New York: Academic Press, 1974).

2. A. W. Crosby, Jr., *The Columbian Exchange: Biological and Cultural Consequences of 1492* (Westport, Connecticut: Greenwood, 1972).

3. J. G. Hawkes, "The History of the Potato," *Journal of the Royal Horticultural Society* 92: 207–24, 249–62, 288–300, 364–65 (1967).

4. J. V. Murra, *The Economic Organization of the Inca State* (Greenwich, Connecticut: JAI, 1980).

5. *Ibid.*

6. E. Hamilton, "American Treasure and the Price Revolution in Spain, 1501–1650," *Harvard Economic Studies* 43: 196 (1934).

7. H. Brucher, *Die Sieben Säulen der Welternährung* (Frankfurt am Main, Germany: Waldemar Kramer, 1982).

8. J. G. Hawkes, "History of the Potato," in P. M. Harris, ed., *The Potato Crop* (London: Chapman and Hall, 1965).

9. Hawkes, "History of the Potato," *Royal Horticultural Society*.

10. G. F. de Oviedo y Valdez, *Historia General y Natural de las Indias* (1526).

11. T. H. Goodspeed, *The Genus Nicotiana* (Waltham, Massachusetts: Chronica Botanica, 1954).

12. N. Monardes, *Historia Medicinal de las Cosas que Sirven al Uso de Medicina* (Seville, 1565).

13. C. Corti, *A History of Smoking* (New York: Harcourt, Brace, 1932).

14. R. F. Smith, "A History of Coffee," in M. N. Clifford and A. C. Wilson, eds., *Coffee, Botany, Biochemistry and Production of Beans and Beverage* (1985).

15. E. F. Robinson, *The Early History of Coffee Houses in England* (London: Kegan Paul, Trench, Trubner, 1893).

16. J., D., and K. Schapira, *The Book of Coffee and Tea* (New York: St. Martin, 1975).

17. Smith, "A History of Coffee."

18. S. J. Stein, *Vassouras: A Brazilian Coffee County, 1850–1900* (New York: Atheneum, 1976).

19. C. W. Bergquist, *Coffee and Conflict in Colombia, 1886–1910* (Durham, North Carolina: Duke University Press, 1978); M. Palacios, *Coffee in Colombia, 1850–1970* (Cambridge, U.K.: Cambridge University Press, 1980); W. Reseberry, *Coffee and Capitalism in the Venezuelan Andes* (Austin: University of Texas Press, 1983).

CHAPTER 10

1. H. J. Habakkuk, "The English Population in the Eighteenth Century," *Economic History Review* 6: 117–118 (1953).

2. A. J. Coale, "The History of the Human Population," *Scientific American* (September 1974).

3. P. Goubert, *The Course of French History* (London and New York: Routledge, 1991).

4. P. Deane and W. A. Cole, *British Economic Growth, 1688–1959* (Cambridge, U.K.: Cambridge University Press, 1969).

5. G. E. Mingay, *The Agricultural Revolution: Changes in Agriculture, 1650–1880* (London: Adam and Charles Black, 1977).

6. D. Marshall, *Eighteenth Century England* (London: Longman, 1974).

7. H. Lee, *The Vegetable Lamb of Tartary* (London: S. Low, Marston, Searle and Rivington, 1887).

8. H. Hobhouse, *Seeds of Change: Five Plants That Transformed Mankind* (London: Sidgwick and Jackson, 1985).

9. E. O. von Lippmann, *Geschichte der Rübe (Beta als Kulturpflanze)* (Berlin, 1925).

10. G. Talbot Griffith, *Population Problems in the Age of Malthus* (1926).

11. P. M. A. Bourke, "The Use of the Potato Crop in Pre-Famine Ireland," *Journal of the Statistical and Social Inquiry Society of Ireland* 12: 72–96 (1968).

12. As mentioned by R. Salaman in *The History and Social Influence of the Potato* (Cambridge, U.K.: Cambridge University Press, 1985), 209.

13. W. W. Cochrane, *The Development of American Agriculture: An Historical Analysis* (Minneapolis: University of Minnesota Press, 1979).

14. C. Sandberg, *Abraham Lincoln: The Prairie Years and the War Years* (New York: Harcourt, Brace, 1954).

CHAPTER 11

1. V. W. Ruttan, H. Binswanger, and Y. Hayami, "Induced Innovation in Agriculture," in C. Bliss and M. Boserup, eds., *Economic Growth and Resources: Natural Resources*, vol. 3, *Proceedings of the Fifth World Congress of the International Economic Association* (London: Macmillan, 1980).

2. World Bank, *World Development Report, 1988* (New York: Oxford University Press, 1988).

3. L. R. Brown and G. W. Finsterbusch, *Man and His Environment: Food* (New York: Harper and Row, 1972).

4. J. F. Richards, "World Environmental History and Economic Development," in W. C. Clark and R. E. Munn, eds., *Sustainable Development of the Biosphere* (Cambridge, U.K.: Cambridge University Press, 1986).

5. National Research Council, *Alternative Agriculture* (Washington, D.C.: National Academy of Sciences Press, 1989).

6. Y. Hayami and V. Ruttan, *Agricultural Development: An International Perspective* (Baltimore: Johns Hopkins University Press, 1971).

7. R. C. Baron, ed., *The Garden and Farm Books of Thomas Jefferson* (Golden, Colorado: Fulcrum, 1987).

8. Brady, *The Nature and Properties of Soils* (New York: Macmillan, 1969).

9. W. Goedert, "Estrategias de Manejo das Savannas," in G. Sarmiento, ed., *Las Sabanas Americanas* (Merida, Venezuela: CIELAT, 1990).

10. N. W. Hudson, "Limiting Degradation Caused by Soil Erosion," in M. G. Wolman and F. G. A. Fournier, *Land Transformation in Agriculture* (Chichester, U.K.: John Wiley, 1987).

11. M. Monteon, *Chile in the Nitrate Era* (Madison: University of Wisconsin Press, 1982).

12. J. R. Kloppenburg, Jr., *First the Seed* (Cambridge, U.K.: Cambridge University Press, 1988).

13. C. B. Huffaker and B. A. Croft, "Integrated Pest Management in the United States," *California Agriculture* 32(2): 6–7 (1978).

14. D. Pimentel, "The Dimensions of the Pesticide Question," in F. H. Borman and S. Kellert, eds., *Ecology, Economics, Ethics* (New Haven: Yale University Press, 1991).

15. As quoted by A. F. Scheuring in "Technological Change and People in Agriculture," *California Agriculture* 34 (1): 6 (1980).

16. National Academy of Sciences, *Genetic Vulnerability of Major Crops* (Washington, D.C.: National Academy of Sciences Press, 1972).

17. This is a relative matter, since food-procuring methods evolved considerably among hunter-gatherers. See B. Meehan and N. White, "Hunter-Gatherer Demography: Past and Present," *Oceania Monograph* 39 (Sydney: University of Sydney, 1990).

18. H. O. Carter, "Agricultural Sustainability: An Overview and Research Assessment," *California Agriculture* 43 (3): 16–18 (1989).

19. National Research Council, *Alternative Agriculture*.

20. T. McKeown, "Food, Infection and Population," in R. I. Rotberg and T. K. Rabb, eds., *Hunger and History* (Cambridge, U.K.: Cambridge University Press, 1983); E. Boserup, *The Conditions of Agricultural Growth: The Economics of Agrarian Change under Population Pressure* (London, 1965), and *Population and Technology* (Oxford: Oxford University Press, 1981).

CHAPTER 12

1. World Resources Institute, *World Resources, 1990–91* (New York: Oxford University Press, 1990).

2. P. R. Crosson and N. J. Rosemberg, "Strategies for Agriculture," *Scientific American* (September 1989): 128–35.

3. M. D. Young, *Sustainable Investment and Resource Use: Equity, Environmental Integrity and Economic Efficiency* (Paris: Parthenon, 1992); M. D. Young and O. T. Solbrig, eds., *Economic Driving Forces and Ecological Constraints in Tropical Savannas* (Paris: Parthenon, 1992).

4. J. Paddock, N. Paddock, and C. Bly, *Soil and Survival: Land Stewardship and the Future of American Agriculture* (San Francisco: Sierra Club, 1986).

5. D. Norse et al., "Agriculture, Land Use and Degradation," in J. C. I. Dooge et al., eds., *An Agenda of Science for Environment and Development into the 21st Century* (Cambridge: Cambridge University Press, 1992).

6. P. S. Fisher, *The Conversion of Agricultural Land to Urban Uses: Towards a Better Model for the Analysis of Public Policy* (Iowa City: Institute of Urban and Regional Research, University of Iowa, 1982).

7. D. Norris, "Environmental Aspects of Agricultural Development," in N. Alexandratos, *World Agriculture: Towards 2000* (London: Belhaven Press, 1990).

8. P. Crosson, "The Issues," in J. Baden, ed., *The Vanishing Farmland Crisis* (Lawrence: University Press of Kansas, 1984).

9. FAO, "Sustainable Development and Management of Land and Water Resources," background document no. 1 for FAO/Netherlands Conference on Agriculture and the Environment, 1991.

10. P. Buringh and R. Dudal, "Agricultural Land Use in Space and Time," in M. G. Wolman and F. G. A. Fournier, eds., *Land Transformation in Agriculture* (Chichester, U.K.: John Wiley, 1987).

11. D. Pearce, "Wells of Conflict on the West Bank," *New Scientist* 130: 36–40 (1953).

12. N. E. Larson, F. J. Pierce, and R. H. Dowdy, "The Threat of Soil Erosion to Long-Term Crop Production," *Science* 227: 458–65 (1983).

13. G. H. Orians and P. Lack, "Arable Lands," in G. A. E. Gall and M. Staton, eds., *Integrating Conservation Biology and Agricultural Production* (University of California Press, 1992).

14. M. B. Holburt, "The Lower Colorado: A Salty River," *California Agriculture* 38 (10): 6–7 (1984).

15. J. R. Harlan, "Our Vanishing Genetic Resources," *Science* 188: 618–21 (1975); J. G. Hawkes, "Conservation of Plant Genetic Resources," *Outlook in Agriculture* 6: 248–53 (1971); National Academy of Sciences, *Genetic Vulnerability of Major Crops* (Washington, D.C.: National Academy of Sciences, 1972).

16. E. Bennett, "The Origin and Importance of Agroecotypes in South-West Asia," in P. H. Davis, P. C. Harper, and I. C. Hedge, eds., *Plant Life in South-West Asia* (Edinburgh: Botanical Society of Edinburgh, 1971).

17. Orians and Lack, "Arable Lands."

18. E. Grafton-Cardwell *et al.*, "Cotton Aphids Have Become Resistant to Commonly Used Pesticides," *California Agriculture* 46 (4): 4–7 (1992).

19. L. Brader, "Plant Protection and Land Transformation," in M. G. Wolman and F. G. A. Fournier, eds., *Land Transformation in Agriculture* (Chichester, U.K.: John Wiley, 1987).

20. M. D. Young, *Sustainable Investment and Resource Use* (Paris: UNESCO, 1992).

21. Board of Agriculture/NRC, *Alternative Agriculture* (Washington, D.C.: National Academy of Sciences, 1989).

22. D. E. Vasey, *An Ecological History of Agriculture* (Ames: Iowa State University Press, 1992).

23. M. A. Young and O. T. Solbrig, "Savanna Management for Ecological Sustainability, Economic Profit and Social Equity," *Mab Digest* 13 (47 pp., 1993).

24. A. Sasson, *Biotechnologies: Challenges and Promises* (Paris: UNESCO,

1984); FCCSET Committee on Life Sciences and Health, *Biotechnology for the 21st Century* (Washington, D.C.: GPO, 1992).

25. C. S. Gasser and R. T. Fraley, "Transgenic Crops," *Scientific American* (June 1992): 62–69.

26. FCCSET Committee, *Biotechnology*, 126.

27. D. Dembo, C. J. Dias, and W. Morehouse, "The Vital Nexus in Biotechnology: The Relationship between Research and Production and Its Implication for Latin America," *Interciencia* 14: 168–80 (1989).

28. M. B. McElroy, "Change in the Natural Environment of the Earth: The Historical Record," in W. C. Clark and R. E. Munn, eds., *Sustainable Development of the Biosphere* (Cambridge, U.K.: Cambridge University Press, 1986); R. W. Corell and P. A. Anderson, eds., *Global Environmental Change* (Berlin: Springer-Verlag, 1991).

29. R. E. Dickinson, "Impact of Human Activities on Climate Ä: A Framework," in Clark and Munn, *Sustainable Development of the Biosphere*; R. E. Dickinson, *The Geophysiology of Amazonia: Vegetation and Climate Interactions* (New York: John Wiley, 1986); G. A. McBean, G. S. Golitsyn, and E. Sanhueza, "Atmosphere and Climate," in J. C. I. Dooge *et al.*, eds., *An Agenda of Science for Environment and Development into the 21st Century* (Cambridge, U.K.: Cambridge University Press, 1992).

30. O. T. Solbrig, "Biodiversity: Scientific Issues and Collaborative Research Proposals," *Mab Digest* 9 (Paris: UNESCO, 1991); M. T. Kalin Arroyo, P. H. Raven, and J. Sarukhan, "Biodiversity," in Dooge *et al.*, *An Agenda of Science*; E. O. Wilson, *The Diversity of Life* (Cambridge: Harvard University Press, 1992).

31. J. Passmore, *Man's Responsibility for Nature* (London: Duckworth, 1974).

Bibliography

Ammerman, A. J., and L. Cavalli-Sforza, 1984. *The Neolithic Transition and the Genetics of Populations in Europe*. Princeton, N.J.: Princeton University Press.

Anderson, E. 1952. *Plants, Life, and Man*. Boston: Little, Brown.

Anderson, E. N. 1988. *The Food of China*. New Haven: Yale University Press.

Bayliss-Smith, T. P. 1982. *The Ecology of Agricultural Systems*. Cambridge, U.K.: Cambridge University Press.

Bender, B. 1975. *Farming in Prehistory*. London: John Baker.

Birrell, V. 1973. *The Textile Arts: A Handbook of Weaving, Braiding, Printing and Other Textile Techniques*. New York: Schocken Books.

Boserup, E. 1965. *The Conditions of Agricultural Growth: The Economics of Agrarian Change under Population Pressure*. London: Earthscan Publications.

Boserup, E. 1981. *Population and Technology*. Oxford: Oxford University Press.

Braudel, F. 1981. *The Structures of Everyday Life (Civilization and Capitalism, 15th–18th Century, vol. 1)*. New York: Harper and Row.

Campbell, B. 1985. *Human Evolution: An Introduction to Man's Adaptation*. 3d ed. New York: Aldine.

Carter, V. G., and T. Dale. 1974. *Topsoil and Civilization*. Norman, Oklahoma: University of Oklahoma Press.

Cochrane, W. W. 1979. *The Development of American Agriculture: An Historical Analysis*. Minneapolis: University of Minnesota Press.

Corell, R. W., and P. A. Anderson, eds. 1991. *Global Environmental Change*. Berlin: Springer-Verlag.

de Vries, J. 1984. *European Urbanization 1500–1800*. Cambridge, Massachusetts: Harvard University Press.

Fagan, B. M. 1986. *People of the Earth: An Introduction to World Prehistory*. Boston: Little, Brown.

Flannery, K. V. 1973. "The Origins of Agriculture." *Annual Reviews of Anthropology* 271–310.

Galloway, J. H. 1989. *The Sugar Cane Industry: An Historical Geography from Its Origins to 1914*. Cambridge, U.K.: Cambridge University Press.

Gras, N. S. B. 1925. *A History of Agriculture in Europe and North America*. New York: S. F. Crofts.

Grigg, D. B. 1974. *The Agricultural Systems of the World: An Evolutionary Approach*. Cambridge, U.K.: Cambridge University Press.

Grigg, D. B. 1982. *The Dynamics of Agricultural Change*. London: Hutchinson.

Harlan, J. 1975. *Crops and Man*. Madison, Wisconsin: American Society of Agronomists.

Hawkes, J. G. 1983. *The Diversity of Crop Plants*. Cambridge, Massachusetts: Harvard University Press.

Hobhouse, H. 1985. *Seeds of Change: Five Plants That Transformed Mankind*. London: Sidgwick and Jackson.

Kalin Arroyo, M. T., P. H. Raven, and J. Sarukhan. 1992. "Biodiversity." In J. C. I. Dooge et al., eds., *An Agenda of Science for Environment and Development into the 21st Century*, pp. 205–19. Cambridge, U.K.: Cambridge University Press.

Kloppenburg, J. R., Jr. 1988. *First the Seed*. Cambridge, U.K.: Cambridge University Press.

Lee, R. B., and E. DeVore, eds. 1968. *Man the Hunter*. Chicago: Aldine.

McElroy, M. B. 1986. "Change in the Natural Environment of the Earth: The Historical Record." In W. C. Clark and R. E. Munn, eds., *Sustainable Development of the Biosphere*, pp. 199–211. Cambridge: Cambridge University Press.

MacNeish, R. S. 1991. *The Origins of Agriculture and Settled Life*. Norman, Oklahoma: University of Oklahoma Press.

Mellaart, J. 1975. *The Neolithic of the Near East*. London: Thames and Hudson.

Mingay, G. E. 1977. *The Agricultural Revolution: Changes in Agriculture 1650–1880*. London: Adam and Charles Black.

National Academy of Sciences. 1972. *Genetic Vulnerability of Major Crops*. Washington, D.C.: National Academy of Sciences Press.

National Research Council. 1989. *Alternative Agriculture*. Washington, D.C.: National Academy of Sciences Press.

Pimentel, D. 1991. "The Dimensions of the Pesticide Question." In F. H. Borman and S. Kellert, eds., *Ecology, Economics, Ethics*, pp. 59–69. New Haven: Yale University Press.

Richards, J. F. 1986. "World Environmental History and Economic Development." In W. C. Clark and R. E. Munn, eds., *Sustainable Development of the Biosphere*, pp. 53–71. Cambridge, U.K.: Cambridge University Press.

Rindos, D. 1984. *The Origins of Agriculture*. Orlando, Florida: Academic Press.

Roosevelt, A. 1987. "The Evolution of Human Subsistence." In M. Harris and E. B. Ross, eds., *Food and Evolution: Toward a Theory of Human Food Habits*, pp. 565–97. Philadelphia: Temple University Press.

Ruttan, V. W., H. Binswanger, and Y. Hayami. 1980. "Induced Innovation in Agriculture." In C. Bliss and M. Boserup, eds., *Economic Growth and Resources: Natural Resources*, vol. 3, pp. 162–89. Proceedings of Fifth World Congress of the International Economic Association. London: Macmillan.

Sahlins, M. D. 1972. *Stone Age Economics*. Chicago: University of Chicago Press.

Sauer, C. O. 1969. *Agricultural Origins and Dispersal*. Cambridge, Massachusetts: MIT Press.

Schwartz, S. B. 1985. *Sugar Plantations in the Formation of Brazilian Society: Bahia 1550–1835*. Cambridge, U.K.: Cambridge University Press.

Solbrig, O. T. 1991. *Biodiversity: Scientific Issues and Collaborative Research Proposals (Mab Digest 9)*. Paris: UNESCO.

Vasey, D. E. 1992. *An Ecological History of Agriculture*. Ames, Iowa: Iowa State University Press.

Wallerstein, I. 1974. *The Modern World System*. 2 vols. New York: Academic Press.

Watson, A. M. 1983. *Agricultural Innovation in the Early Islamic World*. Cambridge, U.K.: Cambridge University Press.

Wilson, E. O. 1992. *The Diversity of Life*. Cambridge, Massachusetts: Harvard University Press.

Young, M. D. 1992. *Sustainable Investment and Resource Use*. Paris: UNESCO.

Zohary, D., and M. Hopf. 1988. *Domestication of Plants in the Old World*. Oxford: Clarendon Press.

Acknowledgments

THIS BOOK WAS originally written as a sourcebook for the use of students in Harvard's core course "Plants and Human Affairs." We wish to thank our students and teaching fellows in that course for the favorable reception of that first effort and for their encouragement, and also thank Dean Susan Lewis and her staff for the help in preparing the sourcebook. A number of friends and colleagues read and commented on this first version. We wish to thank especially Professor Charles Heiser, Jules Janick, Beryl Simpson, Barry Tomlinson, and a number of anonymous reviewers for their useful and insightful comments.

Without the help and good cheer of our editor, Mr. Howard Boyer, this book could not have been published. We would also like to thank Ms. Christine McGowan of Island Press for her efforts, Mr. David Bullen for his design, and very especially Ms. Abigail Rorer for her artistry.

Index